Machining of Nanocomposites

Machining of Nanocomposites

Ramesh Kumar Nayak
Mohan Kumar Pradhan
Ashok Kumar Sahoo

CRC Press is an imprint of the
Taylor & Francis Group, an **informa** business

First edition published 2022
by CRC Press
6000 Broken Sound Parkway NW, Suite 300, Boca Raton, FL 33487-2742

and by CRC Press
2 Park Square, Milton Park, Abingdon, Oxon, OX14 4RN

CRC Press is an imprint of Taylor & Francis Group, LLC

ISBN: 9780367620592 (hbk)
ISBN: 9780367620615 (pbk)
ISBN: 9781003107743 (ebk)

DOI: 10.1201/9781003107743

Typeset in Times
by codeMantra

Contents

Preface

In recent years, the application of composite materials has been increasing in various areas of science and technology due to their unique properties, namely, for use in aircraft, automotive, defense, aerospace, and other advanced industries. The use of nanocomposites for different engineering applications has been increasing day by day. Machining of composite/nanocomposite materials is quite a complex task due to its heterogeneity, and the reinforcements are highly abrasive. This book covers the development of nanocomposite materials and a wide range of conventional and nonconventional machining processes of various composite materials, including polymer and metallic-based composites and nanostructured composites. The machining of polymer matrix composites/nanocomposites is an indispensable reference/sourcebook for composite manufacturing process design, tool, and production engineers. The book also focuses on intelligent and sustainable machining of nanocomposites. It can also benefit final-year undergraduate and postgraduate students as it provides comprehensive information on the machining of composite and nanocomposite materials to produce high-quality final components. It can also serve as a valuable reference work for academics, manufacturing and materials researchers, manufacturing and mechanical engineers, and professionals in composite technology and related industries.

Acknowledgments

The authors take this opportunity to thank Maulana Azad National Institute of Technology, Bhopal, India, an Institute of National Importance, and Kalinga Institute of Industrial Technology (KIIT), deemed to be University, Bhubaneswar, India, an Institution of Eminence' (IoE) for their constant support, motivation, and encouragement in accomplishing the current work. The authors are also thankful to the reviewers, editorial advisory board members, development editor, and CRC Press staff for their time and effort on this project. All of their efforts were crucial in producing this book, and we could not have accomplished this milestone without their constant and consistent advice, support, and collaboration, especially in this COVID-19 pandemic scenario. Finally, we wish to express our deep gratitude to our parents, who always stood by us throughout our lives and guided us in the time of crisis. This project could not be completed without our wives and children's moral support, patience, and encouragement. We express our gratitude to everyone who took the time to assist us in preparing this book. Above all, we bow to the almighty, whose blessings and knowledge have guided and assisted us not only in the course of this project but throughout our lives.

Authors

Dr. Ramesh Kumar Nayak works as Assistant Professor in the Department of Materials and Metallurgical Engineering, Maulana Azad National Institute of Technology, Bhopal, India (an Institute of National Importance). He graduated (MTech) from the Indian Institute of Technology, Kanpur, India, in 2005. He has sound exposure to industries and has worked in reputed organizations such as General Motors India Technical Centre, Bangalore; Hindustan Aeronautics Limited, Bangalore; and DENSO International India Pvt. Limited, Gurgaon, in different capacities. He has been working in developing composite materials and casting technology for the last 16 years. He teaches polymer engineering, solidification and casting, composite materials, and corrosion engineering to undergraduate and graduate scholars at MANIT, Bhopal. He has guided one PhD and several MTech scholars in the area of composite materials. He has published his work in international peer-reviewed journals, reference books, and book chapters and has contributed to the world literature on metallurgy and materials science. His passion is to work closely on industrial problems and develop new technologies/processes to benefit society.

Dr. Mohan Kumar Pradhan, MTech (Mechanical Engineering, REC (now NIT), Rourkela), and PhD (Engineering, NIT Rourkela) is working as Assistant Professor in the Department of Mechanical Engineering, Maulana Azad National Institute of Technology, Bhopal (an Institute of National Importance). He has over 25 years of teaching and research experience in manufacturing. His areas of teaching and research interest are manufacturing, nontraditional machining, metrology, micromachining, MEMS, hybrid machining, composites, and process modeling and optimization. He has 75+ international journal publications; 65+ technically edited papers, which were published in conference proceedings, two edited books, 5+ edited conference proceedings, 5+ journals as guest editor, and 20+ book chapters. He is on the editorial board and review panel of more than five international journals of repute. He has been felicitated with many awards and honors. (Web of Science core collection (193 publications/h-index 16+, SCOPUS/h-index 12+, Google Scholar/h-index 18+). He is charted engineer, life fellow of IIPE, and life member of ISTE, IACSIT, IAENG, and MIE (I).

Dr. Ashok Kumar Sahoo, Professor, Mechanical Engineering, and Director, R&D (Technology), Kalinga Institute of Industrial Technology (KIIT), Deemed to be University, Bhubaneswar, Odisha, India, has 24 years extensive teaching, research, and administrative Experience. He has published more than 150 research articles with h-index 15 (Web of Science), h-index 20 (Scopus), h-index 23 (Google Scholar), i10-index 38 (Google Scholar), with 1290+ and 1710+ Scopus & Google Scholar Citations. Science Direct Top 25 Hottest Articles; five sponsored research projects from DST, SERB, and RPS, AICTE, New Delhi; and eight patents are in his credit. He has supervised 23 MTech and 8 PhD theses. He was the recipient of "Best Staff Award," Kalinga Institute of Industrial Technology (KIIT) University; "Highly Cited Research Award," Elsevier; Outstanding Reviewer award, Elsevier; and reviewer board member in many reputed journals. He is actively engaged in research areas of sustainable machining and machinability of advanced materials, hard machining, MQL and nanofluid-assisted machining, composite development and machinability, condition monitoring, surface characterization, thin film coatings (CVD and PVD), modeling and optimization, etc. He is a fellow of IE(I) and life member of IET (UK), ISTE, and ISCA.

1 Introduction to Nanocomposites and Machining

CONTENTS

DOI: 10.1201/9781003107743-1

1.1 INTRODUCTION

1.1.1 NANOCOMPOSITES

Nanocomposites are materials that incorporate nanosized particles into a matrix of standard material. In polymer matrix nanocomposites, nanoparticles or fillers are embedded. Over the past decade, there has been a growing interest in polymer nanocomposites (PNCs). This is attributed to the improved mechanical, thermal, and electrical properties that enable their use in body parts and structural components of automotive vehicles and aircraft. One drawback of fiber-reinforced polymer (FRP) composites is their weak interface bond between fiber and polymer matrix and brittleness of the polymer matrix. To improve the overall strength of polymer matrix composites, nanofillers are one of the methods researchers and scientists adopt. They have been working to improve FRP composites' strength by adding organic (carbon nanotubes (CNT), single-walled (SWCNT), multiwalled carbon nanotubes (MWCNT), graphene, etc.) and inorganic (Al_2O_3, TiO_2, SiO_2, and ZrO_2, etc.) nanofillers into the polymer matrix. These nanocomposites differ from conventional composite materials due to their high surface-to-volume ratio of the reinforcing and high aspect ratios. There has been growing interest in polymer matrix composites in which nanosized fillers are distributed homogeneously (known as filler/polymer nanocomposites) due to their unique mechanical, electric, optical, and magnetic properties, as well as their thermal and dimensional stability. FRP composites/nanocomposites are used in small and megastructure applications. FRP composite plates or laminates are joined mechanically for large structures.

Therefore, mechanical drilling, abrasive air/water jet machining, milling, laser cutting, electro-discharge machining, etc., are used for the machining of FRP composites/nanocomposites and assembling them with mechanical joints. FRP composites/nanocomposites are anisotropic materials. During machining, there is a chance of fiber delamination, i.e., pull-out from the matrix, leading to degradation of the mechanical properties of the composites. MMCs (metal matrix composites) use metal matrices dispersed with other metal, ceramic, or organic compounds. Reinforcements are usually done to improve the various properties of the base metal. Particle distribution plays a vital role in the properties of MMCs. Copper, magnesium, and aluminum have attracted the most attention as the base metal in metal matrix composites. These MMCs are widely used in aircraft, aerospace, automobile, defense, and various other fields. The most commonly used reinforcements are silicon carbide (SiC), TiO_2, aluminum oxide (Al_2O_3), B_4C, Y_2O_3, Si_3N_4, and AlN. Boron carbide is one of the hardest known elements. It has high fracture toughness and elastic modulus. The addition of boron carbide (B_4C) in MMCs increases their hardness but does not improve wear resistance significantly [1–5].

In recent years, metal matrix nanocomposites (MMNCs) have been widely used in health, transportation, electronics, and aerospace industries [6], which is attributed to their outstanding physical and mechanical characteristics compared to the base metal matrix. Among these, copper-silver nanocomposites can significantly improve the performance of biomedical implants due to their antibacterial effects

[7,8]. The machining of MMNCs is necessary to develop the desired shape and size for different engineering and biomedical applications. Nanofillers are used in FRP composites as filler material to improve nanocomposites' mechanical properties and machinability. The machinability of nanocomposites depends on different machining parameters, and optimizing these parameters can minimize defects during processing. However, challenges to the machinability of nanocomposites still exist, and researchers and scientists are trying to reduce defects produced due to machining of nanocomposites.

Nanocomposites are classified into organic and inorganic nanofiller-enhanced FRP composites. Each type of nanocomposites have their advantages and limitations. However, a common limitation is that with the increase in the addition of wt% of nanofillers in the polymer matrix, the tendency for agglomeration increases. It reduces composites' mechanical properties instead of enhancing them.

This chapter starts with the basic understanding of nanocomposites, their development, characterizations, and application in the real world. It provides brief descriptions of different nanofillers' enhanced polymer matrix nanocomposites. The mixing of nanofillers in the matrix of nanocomposites is a challenge, and different blending or mixing techniques of nanofillers with the polymer matrix are discussed in detail. Nanocomposites undergo a machining process to assemble composite laminates mechanically. Therefore, this chapter is embedded with nanocomposites' development and introduces the conventional and nonconventional machining processes of nanocomposites.

1.1.1.1 Organic Nanofillers

Replacement of FRPs by nanocomposites can be regarded as unrealistic due to the highly developed and well-established conventional fiber-reinforcement of polymers and their still unmatched level of material properties. Nevertheless, the combination of a nanotube-modified matrix together with conventional fiber-reinforcements (e.g. carbon, glass, or aramid fibers), could lead to a new generation of multifunctional materials [9]. Nanoparticles are added to the polymer matrix mechanically, and remain a challenge to realizing the full potential of nano reinforcement. The improvement of materials' performance generally depends on the degree of dispersion, impregnation with matrix, and interfacial adhesion. CNTs are added to the polymer matrix to improve its mechanical properties through interfacial adhesion between CNT and polymer matrix. In FRP composites, the fiber takes the load, and the stress transfer occurs from the polymer matrix to the reinforcement through the interface. Therefore, the chemical functionalization of the CNT surface is necessary for the formation of covalent bonds and additional dipole–dipole interactions between CNTs and the polymeric matrix, resulting in a strengthened interface and an improved wettability of the CNTs [10–12]. The fiber orientation in structural components is usually in the form of a plane (x- and y-direction), leading to fiber-dominated material properties in these directions, whereas the z-direction remains matrix-dominated. The application of CNTs as a reinforcing phase should increase the matrix properties, especially in the z-direction, equivalent to improved interlaminar properties. Recently, Hsiao et al. [13] and Meguid and Sun [14] investigated the tensile and shear strength of nanotube-reinforced polymer composite interfaces by single shear-lap testing. They observed a significant increase in the interfacial shear strength for epoxies with contents between 1 and 5 wt% of multiwall nanotubes compared to the neat epoxy matrix.

1.1.1.2 Inorganic Nanofillers

The interface of matrix and fiber is at the heart of FRP composites. This is because, at the interface, the matrix transfers the load to the fiber. A strong interface bond transfers the load from the matrix to the fiber fully and makes the composites more reliable and durable. Nanoparticles enable the formation good interface bond between polymer matrix and fiber [15]. Therefore, the addition of nanoparticles into the epoxy matrix is one of the methods used to enhance composites' mechanical properties and durability in a hydrothermal environment [16,17]. Inorganic nanoparticles are Al_2O_3, TiO_2, SiO_2, and other metallic and nonmetallic oxides. Therefore, nonmetallic or metallic oxide nanofillers are popular in fiber-reinforced polymer composites due to their low cost and easy fabrication method compared to organic base nanofillers. Inorganic nanofillers are dispersed in the polymer matrix and form physical, mechanical, chemical or a combination of bonds with the epoxy matrix [18]. The well-dispersed nanoparticles in the epoxy matrix enhance the mechanical properties of the nanocomposites. This is because, during the deformation process, the microcrack may divert the path in front of the nanoparticles, blunt or pin on it. As a result, the microcrack needs more energy for the failure to occur [19]. The addition of nano-Al_2O_3 and nano-TiO_2 reduces water diffusivity and enhances hydrothermal durability [3–5,20]. The addition of 2 wt% nanosilica particles into the epoxy polymer increases fracture toughness and wear resistance of the nanocomposites compared to plain epoxy composites in a seawater environment [21].

1.1.2 Mixing/Blending of Nanofillers

Nanofillers are added to the polymer matrix to enhance the mechanical properties of FRP composites. There are various techniques adopted by different researchers and scientists to achieve a homogeneous and uniform distribution of nanofillers in the polymer matrix. However, the proper distribution of nanoparticles in the polymer matrix composites is still a challenge in the field of nanocomposites. This is because nanoparticles have a tendency to agglomerate among themselves and forms clusters, and leading to a decrease in their mechanical, thermal, and electrical properties. The various techniques adopted by different researchers and scientists to achieve uniform distribution of nanoparticle in the polymer matrix are polymerization reaction, surface modification, and chemical reaction of filler materials. On the basis of dimension, these nanofillers can roughly be categorized into one- (Ex. nanographene platelets, montmorillonite clay, layered silicates, etc.), two- (carbon nanotubes, cellulose whiskers, silver/gold nanotubes, etc.), and three- (quantum dots, carbon black, ZnO, Fe_2O_3 and TiO_2 nanocrystals/particles, etc.) dimensional nano-scale fillers, as shown in Figure 1.1 [22]. In this section, some frequently used blending as well as fabrication methods are described.

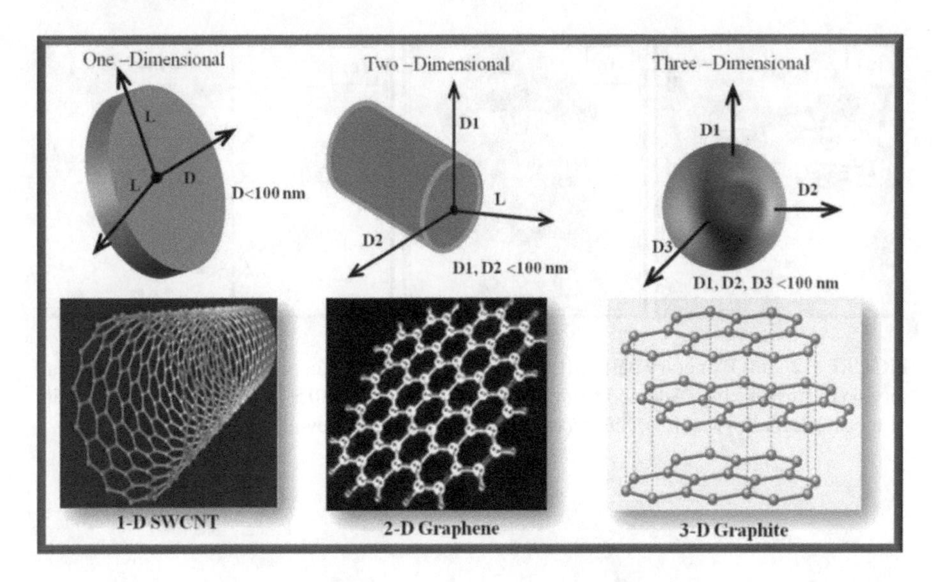

FIGURE 1.1 Classification of nanofillers on the basis of dimension (Adapted from Nayak RK, Ray BC, Rout D, Mahato KK (2020). *Hydrothermal Behavior of Fiber- and Nanomaterial-Reinforced Polymer Composites* (1st ed.). CRC Press [22].)

1.1.2.1 Mechanical Mixing

Ball milling and three-roll mixing are the two methods adopted to mix nanoparticles with polymer matrix. These methods are economical and environmentally friendly to achieve good dispersion of nanoparticles in polymer matrices. In the ball milling process, the mixing occurs at room temperature and is well accepted as a novel green technique, and attracts many researchers to explore the route for preparing advanced functional materials. It follows a top-down approach in which the filler/polymer powders are taken in a container (bowl) at a proper ratio to the solid balls and are closed tight, as shown in Figure 1.2 [23]. This method was successfully applied in the fabrication of carbon-nanofiller-based nanocomposites [24–26].

The three-roll milling, or triple-roll milling, technique is relatively simple, cheap, easy to handle, and scalable; it is a high-volume milling technique widely utilized in industries to mix, refine, homogenize or disperse a range of dense materials like ceramics, carbon/graphite, pigments, plastisol, paints, etc., by making use of a shear force. This shear force results when three horizontally positioned rolls rotate at different speeds and directions from each adjacent roller. A typical three-roll mill and its operation in a schematic diagram is shown in Figure 1.3 [27].

This mixing method is often used to disperse CNTs into the epoxy resin during the fabrication of CNT/epoxy nanocomposites to obtain enhanced electrical, thermal, and mechanical properties [28–31]. Raza et al. employed this technique to disperse graphene in silicone elastomer to improve its mechanical properties [32]. The influence of shear intensity on the property improvement was also investigated in CNT/epoxy-based nanocomposites [30,31].

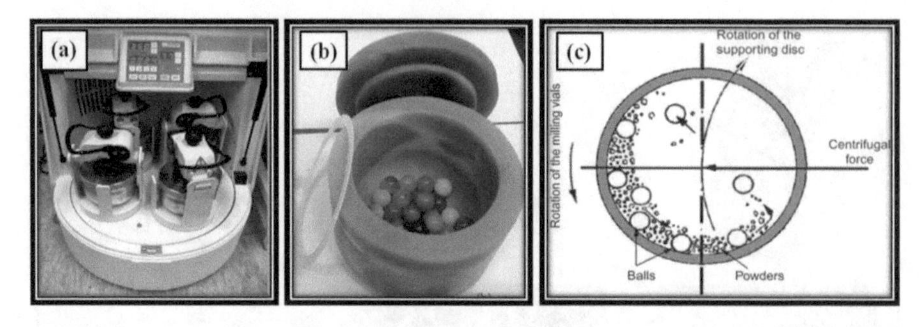

FIGURE 1.2 (a) Planetary ball mill, (b) bowl and balls, and (c) schematic picture of the ball motion during mixing (Adapted from El-Eskandarany, MS (2015). The history and necessity of mechanical alloying, in *Mechanical Alloying* (2nd ed.). Elsevier, pp. 13–47 [23].)

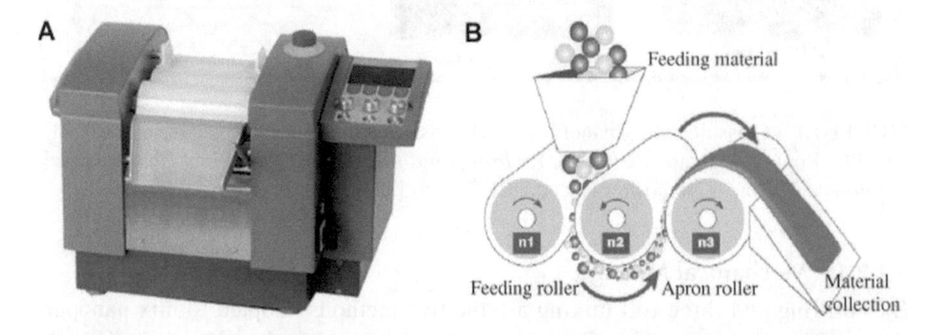

FIGURE 1.3 (a) Calendaring (or three-roll mills) machine used for particle dispersion into a polymer matrix and (b) corresponding schematic showing the general configuration and its working mechanism (Adapted from Ma P-C, Siddiqui NA, Marom G, Kim J-K (2010) Dispersion and functionalization of carbon nanotubes for polymer-based nanocomposites: a review. *Composites Part A: Applied Science and Manufacturing* 41:1345–1367 [27].)

1.1.2.2 Ultrasonic Mixing

Ultrasonic mixing is one of the easiest, cost-effective, and widely used processes for nanoparticle dispersion in a relatively low viscous fluid like water, ethanol, acetone, etc., using a sonicator. In bath sonicators, by principle, ultrasound is generated through a series of attenuated waves via the medium, which assists in breaking agglomerated nanoparticles into individual particles. Commonly, bath sonicators are furnished with an ultrasound frequency of 20 to 23 kHz and an operating power <100 W, but then probe sonicators are equipped with variable amplitudes, between 20% and 70% and a maximum capacity of 1500 W. Sonication is mostly used before the fabrication the final PNC products. It is witnessed that many researchers use the sonication process particularly to break agglomerate CNTs into individual nanotubes before the preparation of CNT-based PNCs [33,34]. Nevertheless, in recent years scientists and researchers employ this technique to disperse CNTs directly into the epoxy matrix with slight modifications to prepare CNT/epoxy nanocomposites [33–36].

1.1.2.3 Magnetic Stirring

Magnetic stirring is one of the most straightforward techniques for the preparation of mixtures, using a magnetic stirrer. This handy laboratory device employs a rotating magnetic field to spin the bar immersed in liquid quickly. Generally, the magnetic field is generated from a rotating magnet (or a set of stationary electromagnets). The vortex effect from the magnet creates a collision between particle-particle along with the bar, resulting in the breaking down of the electrostatic force responsible for particle agglomeration and, thus, produces partial dispersion of the nanoparticles into the polymer matrix. Afolabi et al. have employed magnetic stirring for 2 h at 1250 rpm followed by 90 min sonication to assist good dispersion and interfacial interaction between carbon nanofillers and soy protein matrix [37].

1.1.2.4 Hybrid Mixing

Ball milling, three-roll milling, ultrasonication, and magnetic stirring are widely used to achieve moderate dispersion/exfoliation of various nanoparticles into suitable polymer matrices. However, reports claiming agglomeration/cluster-free dispersion utilizing any one of the above-listed methods individually are rarely available in the literature. Nevertheless, a combination of each of those mixing methods is established as an effective way to eliminate the degree of dispersion and subsequently enhance the properties. Agubra et al. [38] achieved satisfactory clay exfoliation by combining magnetic stirring and three-roll mixing to disperse montmorillonite into epoxy matrix. Ghosh et al. [39] reported that dual ultrasonic mixing (UDM) breaks strong MWCNT agglomerates and assists in achieving a better and homogeneous distribution into the epoxy matrix. Moreover, UDM influences the thermal and tensile properties of MWCNT/epoxy nanocomposites. Similarly, Halder and group reported the influence of UDM on the morphology and thermal and mechanical properties of epoxy/SiO_2 or TiO_2 or ZnO_2 nanocomposites [40–43]. Chun et al. [44] have studied the properties of conducting epoxy/graphene composites prepared by high-speed mechanical stirrer and bath sonicator. Lepico et al. [45] adopted a solution-blending method by combining sonication and magnetic stirring with proper selection of solvents to disperse colloidal nanosilica in polymethylmethacrylate (PMMA).

1.1.2.5 Melt Blending

Melt blending has become a popular technique for preparing PNCs, mostly versatile thermoplastic and elastomeric polymer matrices. In this method, either the polymer is melted first and fillers of the desired amount are added to it or the polymer and filler are dry-mixed and then heated together in an inert atmosphere. Current industrial processes, i.e., extrusion and injection molding, can be categorized under melt blending. Su et al. also obtained encouraging mechanical properties in polycarbonate-graft-graphene oxide composite prepared via melt blending [46]. Similarly, Sharika et al. [47] employed melt blending to design MWCNT-reinforced natural fiber/polypropylene nanocomposite with tunable EMI (electromagnetic interference) shielding performance. However, the main drawback here lies in the poor dispersion of nanofillers into the polymer matrix at higher loadings. Also because of strong shear forces, it may result graphene blocking in the case of graphene-reinforced polymer composites, which can deteriorate the conducting property of the composites [48,49].

1.1.2.6 Functionalization and Grafting

Uniform dispersion of nanoparticles of suitable surface morphology within the polymer matrix is highly challenging in the design and fabrication of composites with novel macroscopic properties. In order to develop requisite physicochemical properties, control/modification of surface morphology of the nanofillers is desirable, utilizing their large surface to volume aspect ratio and strong surface effects. Hence, surface engineering of nanofillers is essential for the deliberate manipulation of the properties of PNCs. In this course of achieving homogeneous dispersion and subsequently enhanced interfacial interactions, the functionalization of nanofillers and grafting of polymers play a vital role. Generally, functionalization is of two types: chemical (covalent) and physical (noncovalent). A few of the possibilities of functionalization of CNTs are shown in Figure 1.4a [50]. Figure 1.4b shows a schematic diagram of the grafting-from and grafting-to methods [51]. For surface grafting, several methods such as UV irradiation, plasma discharge, etc., are successfully used by researchers. Wei et al. have made PNCs with polymer-grafted nanoparticles, which exhibit better crystallinity, ductility, Young's modulus, and tensile strength [52].

1.1.2.7 Solution Mixing

Solution mixing is one of the most common, easy, and efficient methods used for the synthesis/fabrication of PNCs accompanying a wide range of polymers such as polyethylene, polyvinyl fluoride, polyvinyl alcohol, PMMA, epoxy, polyurethane, etc. Moreover, the technique requires no expensive equipment and operative protocols despite quite simple steps such as dispersion of fillers, incorporation of the polymer, and removal of the solvent. A schematic diagram of this technique is shown in Figure 1.5 [53].

Djahnit et al. [54] obtained thermally stable Zinc oxide/PMMA nanocomposite with UV shielding capability and optical transparency. Gong et al. [55] fabricated temperature-independent piezoresistive sensors utilizing CNT/epoxy resin

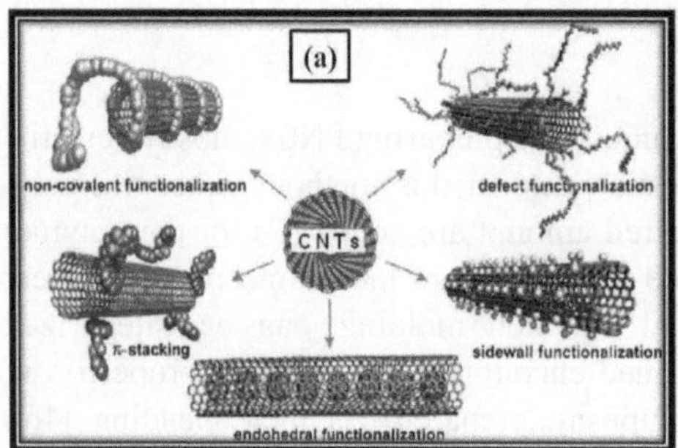

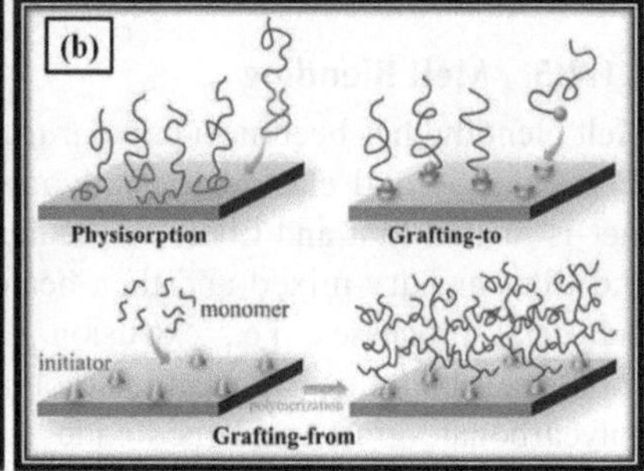

FIGURE 1.4 (a) A few possibilities of functionalization of CNT and (b) a schematic diagram of the grafting-from and grafting-to methods. (Adapted from Hussain CM, Mitra S (2011) Micropreconcentration units based on carbon nanotubes (CNT). *Analytical and Bioanalytical Chemistry* 399:75–89 [50,51].)

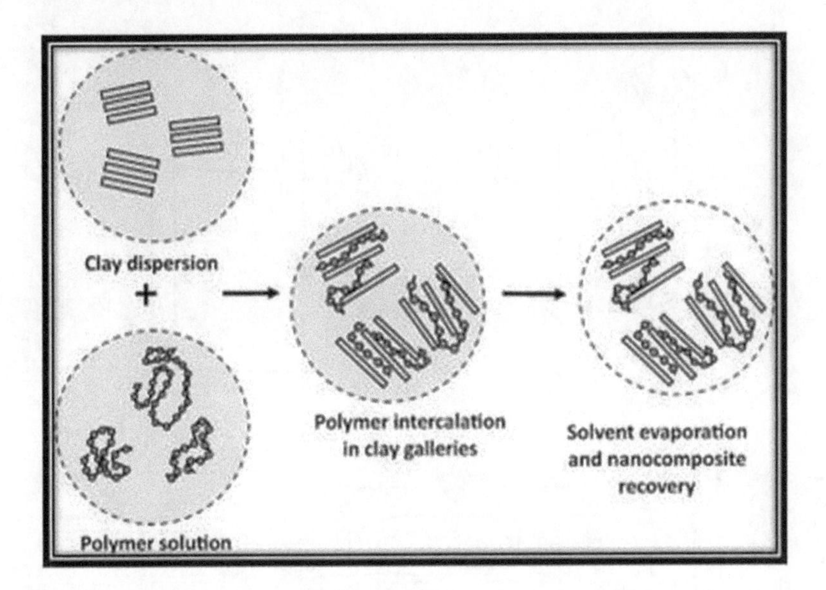

FIGURE 1.5 Synthesis of nanocomposites by dispersion in solution mixing. (Adapted from Mtibe A et al. (2018) Fabrication and characterization of various engineered nanomaterials, in *Handbook of Nanomaterials for Industrial Applications*. Elsevier, pp. 151–171 [53].)

nanocomposites prepared through the solution method. However, the use of expensive solvents and their disposal creates a hindrance to scaling up and for their eventual adaptation by industries. Nevertheless, distilled water is used as a solvent to prepare some carbon nanofiller/rubber nanocomposites [56,57].

1.1.3 Fabrication of Nanocomposites

The utility of PNCs in various sectors has been increasing, and many preparation methods are developed and employed to fabricate numerous PNCs. PNC fabrication is a complex process, and it is integrated by the factors of design, manufacturing, and economics. It requires concurrent thought over various parameters such as reinforcement and matrix types and their compatibility, component geometry and structure, production scale or volume, tooling or equipment requirements, environmental aspects, and process economy. Some of the fabrication techniques of PNCs are discussed here.

1.1.3.1 Hand Lay-Up Method

Hand lay-up (often called wet lay-up) is the primitive, typical, simple, and least expensive open mold method suitable for fabricating a wide variety of polymer composites expanding from small and large-scale productions using low-cost equipment. Despite low production volume per mold, it is feasible to yield a substantive amount utilizing multiple molds. Figure 1.6 shows a schematic picture of the hand lay-up equipment [58].

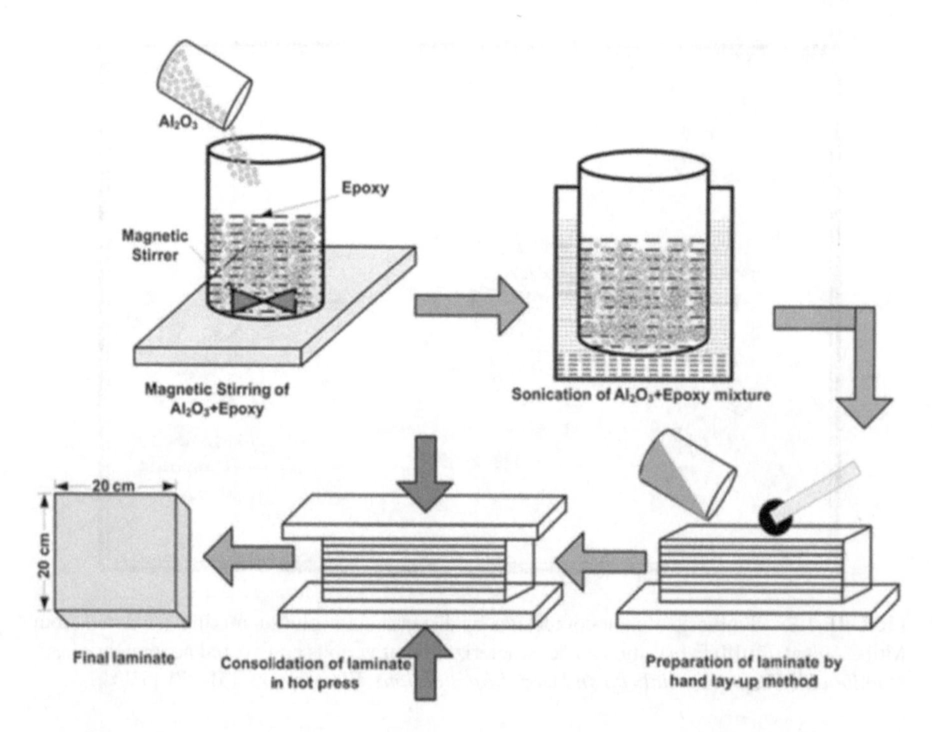

FIGURE 1.6 A schematic picture of the hand lay-up method. (Adapted from Nayak RK (2019) Influence of seawater aging on mechanical properties of nano-Al2O3 embedded glass fiber reinforced PNCs. *Construction and Building Materials* 221:12–19 [58].)

Hallonet et al. [59] recently prepared flax/epoxy composites suitable for the use of external strengthening of RC structures. Seretis et al. [60] investigated post-curing effect on the thermal and mechanical behavior of graphene-nanoparticle-reinforced hand lay-up glass fabric/epoxy nanocomposites. Similarly, Budelmann et al. [61] reported the influence of process (temperature, lay-up, and compaction speed) and materials-related (resin, age, draping/lay-up surface) parameters on material properties with regard to the properties of epoxy-impregnated carbon fibers.

1.1.3.2 Vacuum Resin Transfer Molding

The vacuum-assisted resin transfer molding (commonly known as VARTM) process has been established as one of the most reliable, low-cost, and widely used methods for fabricating FRP composites over the two decades since its development. This technique differs from prepreg laminate composite methods in which the resin is infused into the dry fabric on a mold to produce near net shape products under vacuum pressure and cured in an oven. A schematic diagram of a typical VARTM method is shown in Figure 1.7 [62].

Besides, the method has a list of advantages over the other processes, i.e., (i) low emission of volatile organic compounds, (ii) high product quality, (iii) repeatability, (iv) affordability, (v) clean handling of RTM process, (vi) scalability of open mold hand lay-up process, (vii) flexible mold tooling design, and so on. However, the

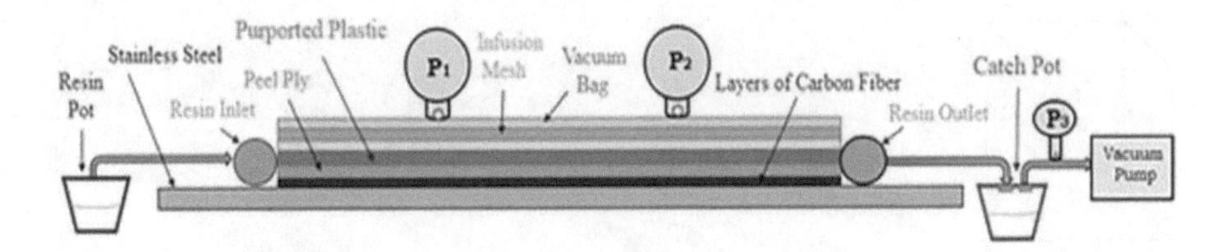

FIGURE 1.7 A typical VARTM method. (Adapted from Tamakuwala VR (2021) Manufacturing of fiber reinforced polymer by using VARTM process: A review. *Materials Today: Proceedings* 44:987–993 [62].)

technique has limitations: the flow distribution medium, vacuum bag, sealing tape, peel ply, and resin tubing may not be reusable; high degree air leakage; limited resin injection pressure; and compression. Kong et al. [63] designed and manufactured automobile hood for structural safety and stability using flax/vinyl ester composite. The group has also studied the effect of ZnO nanoparticles on the mechanical properties of ZnO/polyester woven carbon-fiber composite fabrication using this method [64]. Nisha et al. [65] have prepared electrostatic-bonded polyvinylidene fluoride (PVDF)-MWCNT and glass-fiber-reinforced polymer composites for structural health monitoring.

1.1.3.3 Filament Winding

Filament winding is the oldest manufacturing process for composites, used since the early 1960s and 1970s to fabricate fiber-reinforced pipes, streetlight poles, and pressure vessels. This technique has progressively turned out as a fast, cost-effective method to create lightweight, high-performance structures, including driveshafts and golf club, oars/paddles, yacht masts, small aircraft fuselages, bicycle rims and forks, car wheels and pressure vessels, spacecraft structures, cryogenic fuel tanks for spacecraft, liquid propane gas tanks, and firefighter oxygen bottles owing to its suitability to automation and the escalated use of robotics and digital technologies. A typical filament winding machine is shown in Figure 1.8 [66].

Nevertheless, there is still ample scope to improve this technique to address some of the drawbacks: (i) the method is preferably used only for convex-shaped components, (ii) the high cost of mandrel for more complex components, (iii) the external surface of the components are cosmetically unattractive, and (iv) high-viscosity resins create health and safety concerns. Zhao et al. [67] have used this technique to design shock-less smart releasing shape memory composites.

1.1.3.4 Pultrusion

Pultrusion is a speedy, systematic, economic, and continuous manufacturing process for products with a constant cross section, e.g., structural shapes, rod stock, pipe, beams, channels, gratings, tubing, golf club shafts, decking, fishing rods, and cross arms. This continuous molding process uses polymers to form high-strength structural parts of a constant length and makes fiber-reinforced PNCs. Fundamentally, it comprises two steps: (i) forming process and (ii) heat treatments to produce uniform cross sections/lengths. The reinforcements, e.g., continuous-strand glass fiber, basalt

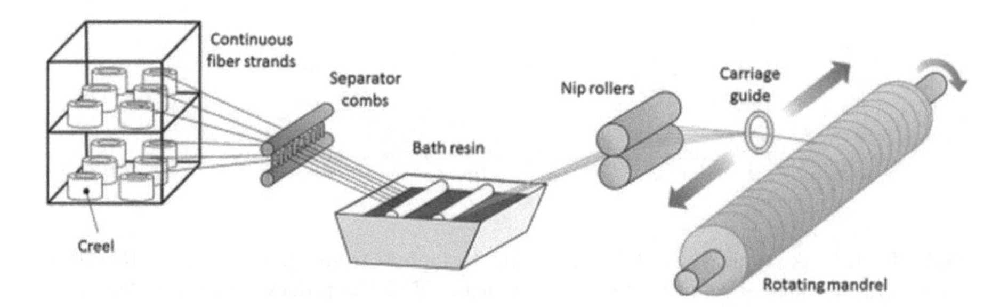

FIGURE 1.8 A typical filament winding machine (Adapted from Shrigandhi GD, Kothavale BS (2021) Biodegradable composites for filament winding process. *Materials Today: Proceedings* 42:2762–2768 [66].)

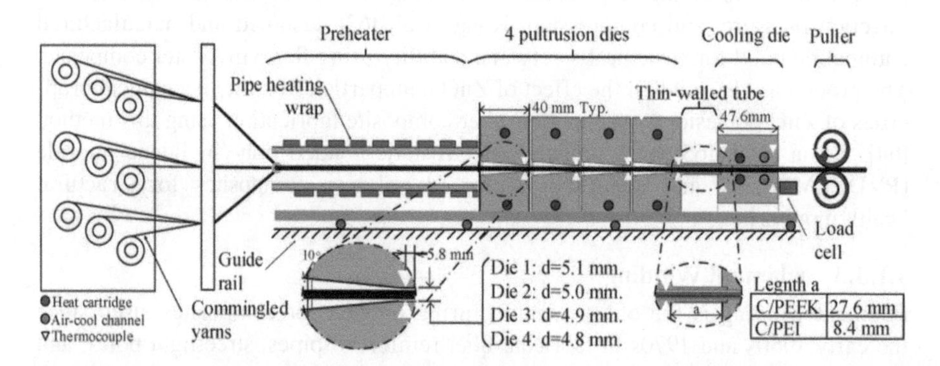

FIGURE 1.9 Schematic of the multi-die pultrusion system (The figure was reproduced from Alsinani N, Ghaedsharaf M, Laberge Lebel L (2021) Effect of cooling temperature on deconsolidation and pulling forces in a thermoplastic pultrusion process. *Composites Part B: Engineering* 219:108889 [68].)

fiber roving, carbon fiber, mat, and surfacing veil, are generally pulled (by a pulling system) through the impregnation system (resin bath and performers: series of steel dies), as shown in Figure 1.9 [68].

Bowlby et al. [69] developed some biocomposites by reinforcing carbon-based biochar particles in pultruded glass FRP composites with improved flexural strength. Similarly, Chang et al. [70] have studied the mechanical behavior of heat-treated pultruded-kenaf fiber-reinforced polyester composites. Recently Saenz-Dominguez's group has designed cellular composite structures via out-of-die UV-cured pultrusion for automotive crash-boxes [71].

1.1.4 Machining of Nanocomposites

1.1.4.1 Drilling

Fiber-reinforced polymer composites and nanocomposites are used in different industries due to their advantages over metallic parts. They are used in megastructure and structural applications, such as in airports, marine, and in

FIGURE 1.10 Drilling setup (The figure was reproduced from the referenced from Kumar J, Kumar Verma R, Debnath K (2020) A new approach to control the delamination and thrust force during drilling of PNCs reinforced by graphene oxide/carbon fiber. *Composite Structures* 253:112786 [72].)

organizing Olympics events. Although composite manufacturers avoid multiple joints in megastructure applications, one composite cannot be sufficient for use in megastructures. Therefore, FRP composites must be machined to assemble different composites into a big frame. In the manufacturing process, the primary machining process is drilling, and this is required to produce two composite laminates through rivets and fastening of various parts to one end product.

During drilling, the hole dimension should be precise; imprecise holes result in improperly assembled structures and reduce product durability and efficiency. The drilling procedure is frequently used to create an intricately exact hole in the material to be used in different manufacturing applications [72]. The machining of FRP composites is quite distinct from metals due to their capacity for plastic deformation and their inhomogeneous structure [73]. Conventional machining, such as computer numerical control (CNC) drilling, is used to machine polymer to produce an intricate hole with the desired surface quality. The drilling setup for FRP machining is shown in Figure 1.10 [74].

The details of the working principle are discussed in the subsequent chapters. Delamination of composites during machining is very critical, and lower thrust force may avoid damages. Delamination at the entry and exit points of the hole can lead to poor dimensional stability and imbalance the force or drill bit. Therefore, desired process parameters are necessary for excellent machining of FRP composites. Different scientists and researchers have used experiments and optimization techniques to optimize the process parameters for minimum delamination and surface roughness. Kumar et al. [74] studied the effect of wt.% of graphene oxide (GO), spindle speeds, and feed rate on the quality of the drilled sample through response surface methodology (RSM) and optimized the input parameters.

1.1.4.2 Milling

Nanofillers such as graphene oxides, carbon nanotubes, and nano-ceramic oxides are added to polymer matrix composites to enhance their physical, mechanical, thermal, and electrical properties. To improve the dimensional accuracy and better surface finish of composite laminates without disturbing the alignment of the particulate, mechanical machining processes such as milling are used in industry. Thus, to enhance the performance of these new polymeric nanocomposites' performance, the material removal behavior of the composite needs to be well understood. Figure 1.11 shows a schematic representation of the milling operation of nanocomposites [75]. The details of the working principle are discussed in the subsequent chapters. They have performed milling operations on nanocomposite specimens to understand material removal and chip morphology. The effect of cutting forces on material removal of graphene nanoplatelet (GNP)-reinforced composites and plain polycarbonate (PC) mat was studied, and it was found that GNP-filled PC nanocomposites require higher cutting force than plain PC. It is observed that the addition of GNPs and MWCNTs can significantly improve the dimensional accuracy and the quality of the machined surfaces. Samuel et al. [76] investigated the machinability of CNT/PC composites. They experimentally showed that cutting forces were less for CNT-based than those with base PC or carbon-fiber-loaded PC. Dikshit et al. [77] adopted a microstructure-based machining simulation model for the machining of CNT nanocomposites. They modeled the PC phase by considering its behavior at significant strain and included the thermal effect, while the CNTs were modeled as a perfectly elastic material.

Arora et al. [78] explored the machinability of epoxy-reinforced graphene platelets with small loadings (i.e., 0.1–0.3 wt%). They observed higher resultant cutting forces for 0.2 wt% graphene-filled PC composites than the base PC, which was attributed to the improved mechanical properties of the nanocomposite at this nanofiller loading. Nasr et al. [79] studied the effect of graphene reinforced in Ti6Al4V alloy on

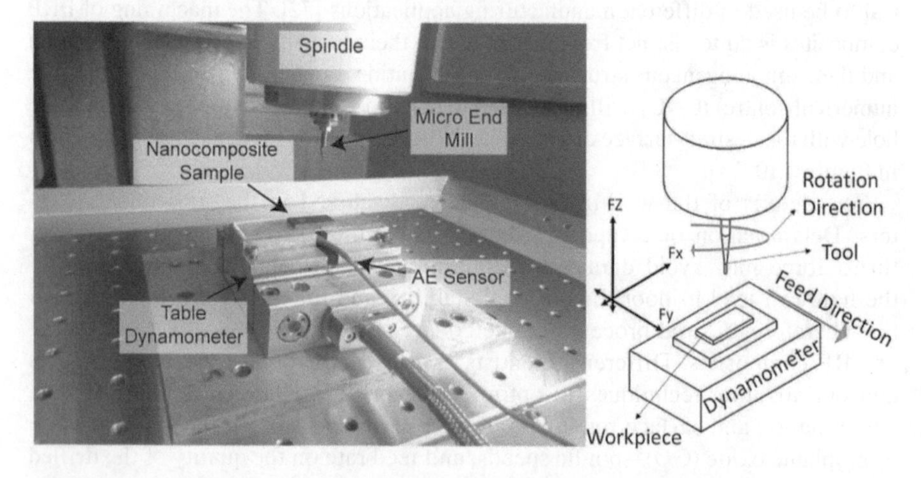

FIGURE 1.11 Schematic representation of milling operation of nanocomposites (The figure is adapted from Kumar J et al. (2020) *Composite Structures* 253:112786 [75].)

milling machining parameters. They investigated the effect of cutting speed, feed rate, and GNP ratio on the machining behavior of nanocomposites. They found that the multi-objective optimization results from the new approach indicated that a cutting speed of 62 m/min, a feed rate of 139 mm/min, and a GNPs reinforcement ratio of 1.145 wt% lead to the improved machining characteristics of GNPs-reinforced Ti6Al4V matrix nanocomposites.

1.1.4.3 Abrasive Air-Jet Machining

During mechanical assembling of composite components, drilling is required to make holes in the composites. The machining of FRP composites by conventional methods often leads to failure of the composite assembly due to different damage mechanisms, such as fiber pull-out, delamination, and matrix cracking. The drilling operation is generally performed to assemble parts/components into the final shape with rivets and bolts. The machining of FRP composites by conventional cutting tools can be replaced by alternative, nonconventional machining techniques such as abrasive water-jet machining (AWJM) and abrasive air-jet machining (AAJM). Figure 1.12 shows a schematic diagram of AWJM [80].

The details of the working principle are discussed in the subsequent chapters. In AAJM, the operating air pressure is much lower, at about 3–10 bar. Rao et al. [81] performed optimization of the material removal rate (MRR) of glass-fiber-reinforced epoxy composite by AAJM using response surface methodology. They obtained the

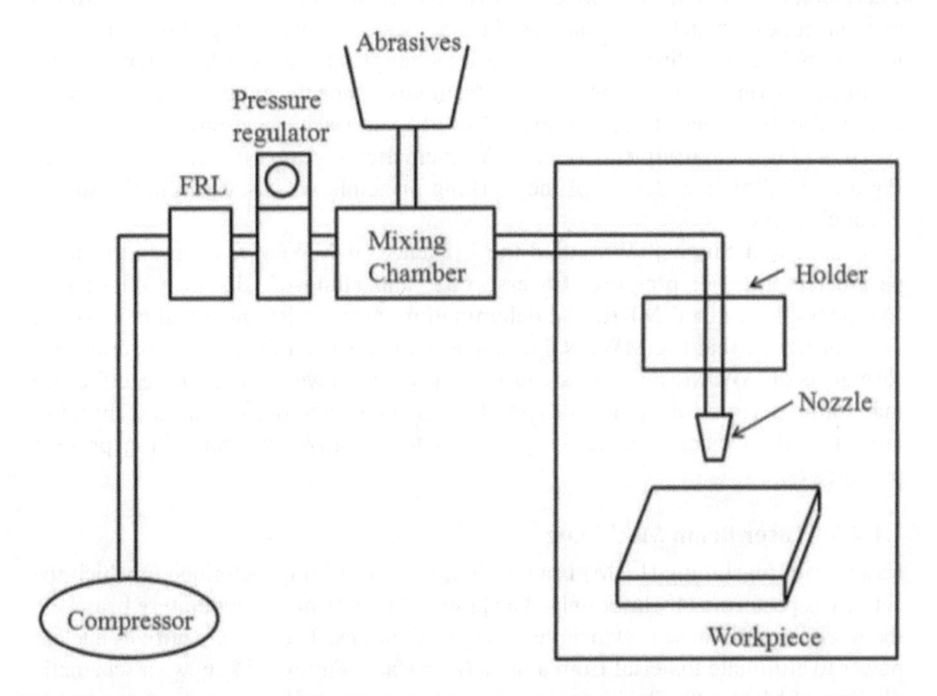

FIGURE 1.12 Schematic diagram of AAJM (Adapted from Jesthi DK, Nayak RK (2020) Sensitivity analysis of abrasive air-jet machining parameters on machinability of carbon and glass fiber reinforced hybrid composites. *Materials Today Communications* 25:101624 [80].)

highest MRR at the operating pressure of 8 kg/cm^2 using a 4-mm-diameter nozzle with a 10 mm stand-off distance.

Sharma and Deol [82] machined glass-fiber-reinforced plastic by AAJM using silicon carbide as the abrasive material. They observed that the MRR increases with increasing pressure and nozzle diameter. However, MRR decreases with an increase in stand-off distance. Jesthi and Nayak [83] investigated the sensitivity of AAJM process parameters on MRR, depth of cut (DOC), and surface roughness (Ra), and optimized them through RSM. The optimum machining process output predicted by the RSM model for a [CG2CG]S-type hybrid composite is as follows: MRR = 3.26×10^{-3} g/sec, DOC = 1.48 mm, and surface roughness (Ra) = 1.48 μm. Madhu and Balasubramanian [84] carried out AAJM of carbon-fiber-reinforced polymer (CFRP) composites using an internally threaded nozzle. They reported better surface roughness of composite compared to the regular unthreaded nozzle. Surface roughness is improved by increasing the pressure and decreasing the stand-off distance.

1.1.4.4 Abrasive Water-Jet Machining

Conventional machining of FRP composites generates heat, and the matrix melts or burns at that temperature. Besides heat, the cutting forces create various damages, such as filer pull-out, tool wear, delamination, and rough surface finish due to nonhomogeneous and anisotropic characteristics of the composites [85]. Therefore, there is a need to machine FRP composites or nanocomposites through nonconventional machining, such as abrasive water jet machining. AWJM is being widely applied in industries among these techniques because it has a few distinct advantages over other nonconventional techniques. For instance, it does not create a heat-affected zone, it ensures high flexibility and accuracy and higher cutting speed, and it is environment-friendly. A schematic diagram of AWJM is shown in Figure 1.13 [86]. The details of the working principle are discussed in the subsequent chapters.

Thakur and Singh [87] studied the influence of MWCNT wt.% and process parameters like jet pressure, traverse rate, and stand-off distance on surface roughness, kerf taper, MRR, and delamination factor at jet entry and exit points. They observed that the MWCNT wt% and jet pressure influence the overall performance of AWJM, and the stand-off distance showed a negligible effect on machining performances. The design of experiments and optimization techniques are adapted to machining nanocomposites to optimize the machining process parameters.

1.1.4.5 Laser Beam Machining

Laser beam machining (LBM) is a non-ordinary machining technique in which the activity is performed by laser light. The laser light emits high-temperature beams on the workpiece, because of which the workpiece softens. The process utilizes nuclear power to eliminate material from a metallic surface. Figure 1.14 shows a schematic diagram of LBM [88]. The high frequency monochromatic light falls on a superficial level and heating, melting, and disintegrating of the material occur because of impinging of photons. LMB is most appropriate for weak materials with low conductivity,

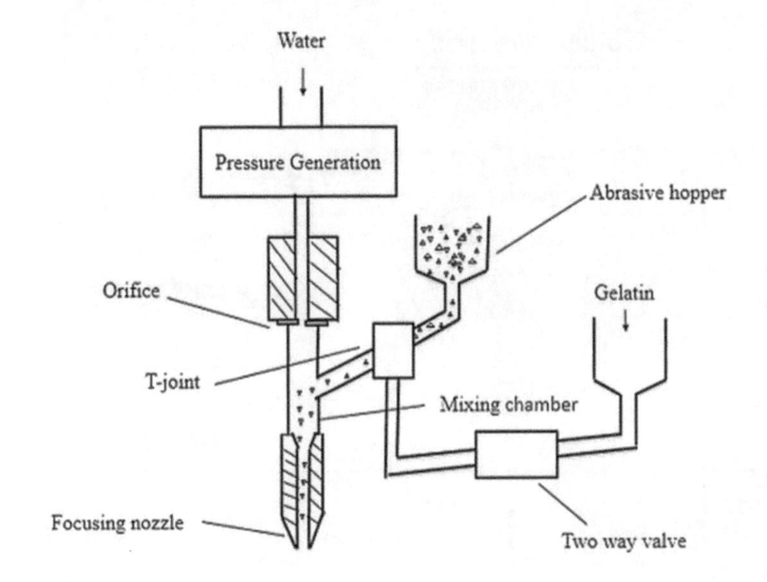

(a)

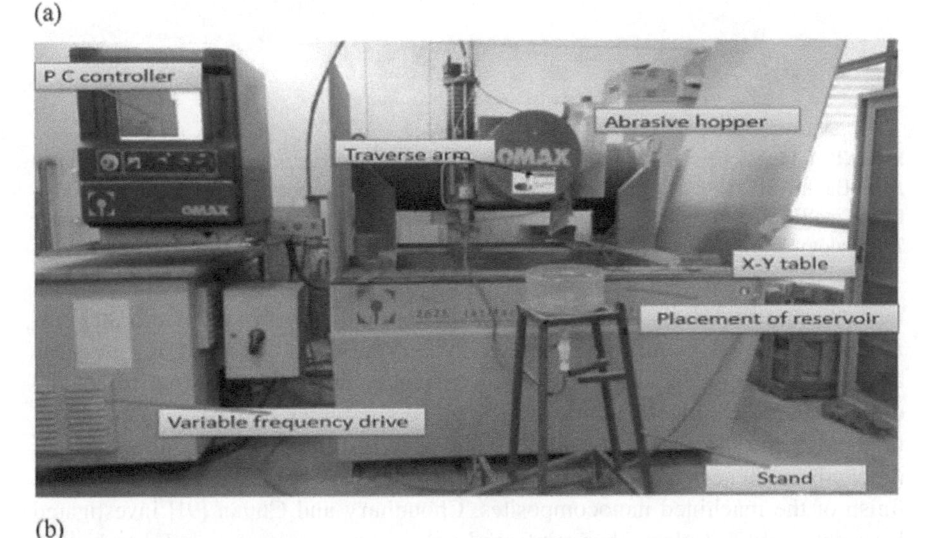

(b)

FIGURE 1.13 (a) Schematic diagram of abrasive water slurry jet machine and (b) the experimental AWSJM setup (Adapted from Amar AK, Tandon P (2021) Investigation of gelatin enabled abrasive water slurry jet machining (AWSJM). *CIRP Journal of Manufacturing Science and Technology* 33:1–14 [86].)

yet can be utilized on most materials [89]. The effect of parameters such as laser power, scanning speed, pulse frequency, and hole diameter has been investigated on the quality of cut, such as taper, heat-affected zone (HAZ), kerf width, and surface roughness, by adopting CO_2 and Nd:YAG lasers.

Nagesh et al. [90] studied the effect of Nickel nanopowder and carbon black nanofillers on laser beam machinability of glass-fiber-reinforced vinyl ester

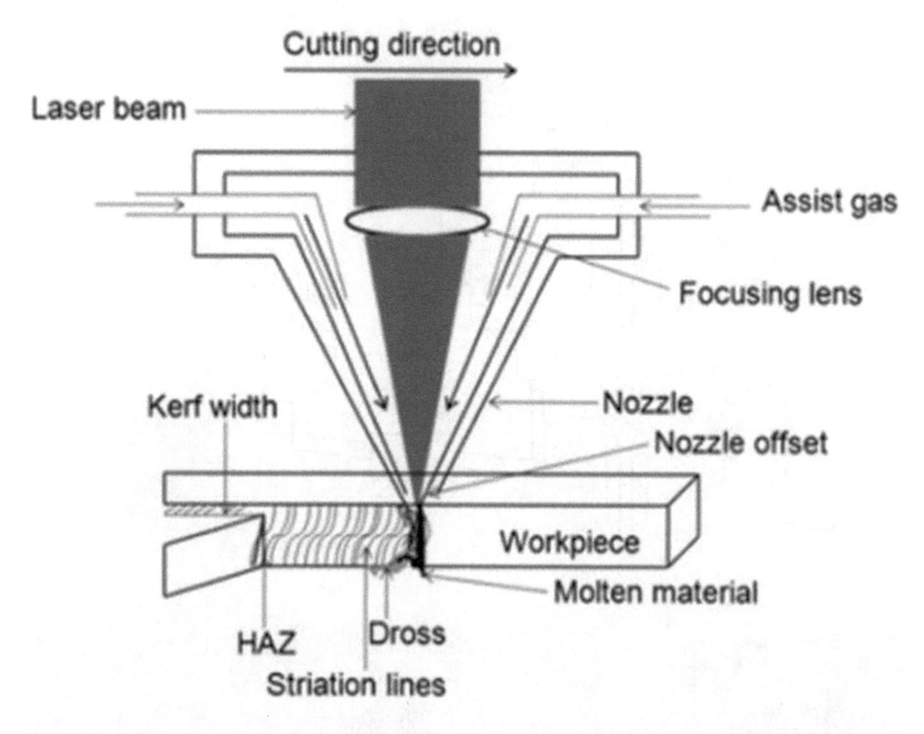

FIGURE 1.14 Schematic diagram of LBM. (The figure has reproduced from the referenced from Bakhtiyari AN, Wang Z, Wang L, Zheng H (2021) A review on applications of artificial intelligence in modeling and optimization of laser beam machining. *Optics & Laser Technology* 135:106721 [88].)

matrix nanocomposites. They analyzed the top and bottom kerf width and surface roughness of the machined surface by varying scanning speed, air pressure, and nanofillers content at a constant power source. They adapted experiments and optimization techniques to optimize the process parameters to achieve good surface finish and dimensional stability. They observed that the scanning speed significantly influences machinability and that nanofillers help achieve good surface finish of the machined nanocomposites. Choudhury and Chuan [91] investigated laser cut quality of glass-fiber-reinforced polymer by single-pass and double-pass laser beams experimentally using first-order orthogonal design. The results show that the quality of cut surface produced by a double-pass laser beam appeared to be much better than that produced by a single-pass laser beam. Tagliaferri et al. [92] studied the machined surface of 500 W CO_2 laser cut of aramid, graphite, and glass-fiber-reinforced polymer composites. They observed an excellent surface finish and low HAZ of the machined composites. Yung et al. [93] studied laser drilling performance of blind vias in epoxy/aluminum nitride (AlN) composites using Nd:YAG laser and found that using a higher thermal conductivity filler, such as AlN, reduces the HAZ. Karippal et al. [94] reported that dispersion of nanofillers (MWCNT and carbon black) resulted in an increase in thermal conductivity of the polymer and improved its machinability.

1.1.4.6 Ultrasonic Machining

A metal tool is given a high-frequency, low-amplitude oscillation perpendicular to the workpiece, which transmits a high velocity to fine abrasive particles present between the tool and the workpiece. These particles strike the workpiece, chipping away small particles, and the tool is gradually fed into the workpiece. The "chips" are carried away from the workpiece by a constant flow of cooled slurry. The workpiece is abraded into a mirror image of the tool. In rotary ultrasonic machining (RUSM), the tool is rotated as well as being vibrated. This type of machining is most commonly used for drilling using diamond-plated drills, which has greater accuracy and cutting speed. The general principle of ultrasonic-assisted machining (UAM) is applying high frequency (20–40 kHz) and low peak-to-peak amplitude (up to 10 μm) to the tool or workpiece. Figure 1.15 shows a schematic diagram of (a) UAM and (b) details of machining movements in diamond tool–zirconia sample contact [95].

Machining CFRP composites is challenging as the anisotropic and nonhomogeneous nature of CFRP makes it difficult to machine. The highly abrasive nature of carbon fibers leads to rapid tool wear and reduced tool life. Delamination, fiber and matrix pull out, matrix cracking and smearing, and rapid tool wear are the most common problems reported when machining CFRP using the conventional machining process. Huda et al. [96] studied tool condition, cutting forces, and surface integrity of the machined surface of CFRP composites. They used a tool with a diameter of 10 mm, and a nickel-bonded, 420 μm diamond grit coating was used with a constant speed of 565 m/min, a feed rate of 1500 mm/min, and radial depth of cut of 3 mm.

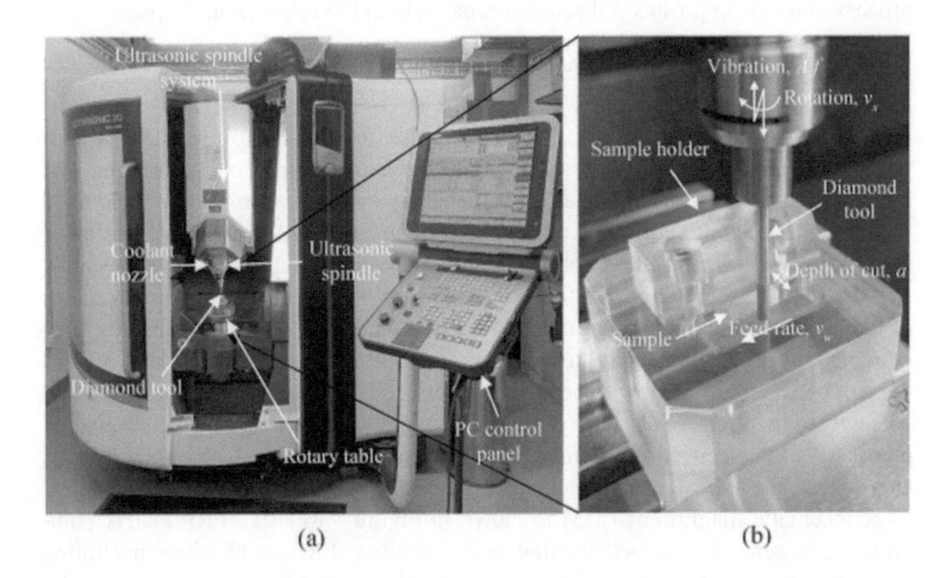

FIGURE 1.15 Schematic diagram of (a) ultrasonic vibration-assisted machining and (b) details of machining movements in diamond tool–zirconia sample contact (Adapted from Juri AZ, Nakanishi Y, Yin L (2021) Microstructural influence on damage-induced zirconia surface asperities produced by conventional and ultrasonic vibration-assisted diamond machining. *Ceramics International* 47:25744–25754 [95].)

A frequency of 43 kHz and an amplitude of 6.9 μm were employed. They found a reduction in cutting force, better tool condition, and improved surface roughness compared to conventional machining. Significant wear occurring in conventional machining is grain pull-out and wear flats of the diamond grit.

In contrast to UAM, microcracks in the diamond grits were observed after a 10 m machining length, suggesting that UAM can prolong tool life. The cutting forces and surface roughness recorded by UAM showed 20% improvement compared to conventional machining. Majeed et al. [97] performed drilling operations on composite Al_2O_3 /$LaPO_4$ through ultrasonic machining using low-carbon steel tools. Their results indicated the significance of adding $LaPO_4$ in the MRR, acoustic emission, and geometry of the drilled hole. Moreover, they observed that the tool–workpiece contact ratio (TWCR) is a significant factor in evaluating the cutting speed and the tool wear. Liu et al. [98] determined the optimum condition of process parameters of ultrasonic vibration machining on SiC monocrystals. Singh and Gianender [99] discussed the effect of process parameters on MRR and surface roughness for the ultrasonic machining of hard and brittle materials, such as ceramics, glass, and composites. Chakravorty et al. [100] compared sets of past experiment data to the recently obtained experiment response of ultrasonic machining to optimize process parameters. Zhou et al. [101] reported Al/45% SiC composite machinability by rotary ultrasonic face grinding. They examined the effect of ultrasonic vibration on cutting force, abrasive chip shape, tool wear, and surface quality. Tool wear was mainly due to the grain breakage and grain fall-off but not the observed grinding wheel blockages and grinding burn. A detailed understanding of this nonconventional machining process of nanocomposites is discussed separately in the subsequent chapters.

1.1.4.7 Electrical Discharge Machining

A potential difference in the electrical discharge machining (EDM) process is applied across the tool and workpiece in pulse form. The tool and workpiece must be electrically conductive, and a small gap is maintained between them. The tool and workpiece are immersed in a dielectric medium (kerosene or deionized water). As the potential difference is applied, electrons flow from the tool and move toward the workpiece. In this process, the tool is negative and the workpiece is positive. The electrons moving from the tool to the workpiece collide with the molecules of the dielectric medium. Due to the collision of electrons with the molecules, they get converted into ions. This increases the concentration of electrons and ions in the gap between tool and workpiece. The electrons move toward the workpiece and ions toward the tool. An electric current is set up between tool and workpiece, called plasma. As the electrons and ions strike the workpiece and tool, electrons' kinetic energy changes to heat energy, and machining starts.

A schematic diagram of EDM is shown in Figure 1.16 [102]. FRP matrix composites are generally non-conducting and, therefore, FRP composites machining is not as simple as for metallic materials. Chaudhury and Samantaray presented a review article on the role of CNTs for enhancing surface quality through EDM. They stated that as many authors have studied the feasibility of machining CNT composites through EDM, the performance of such a machining process is higher in terms of surface finish and controlled MRR, which will be helpful for the use of CNTs in

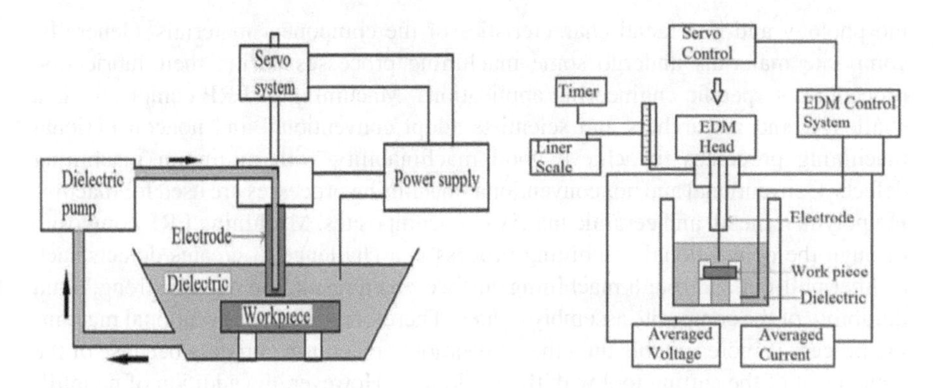

FIGURE 1.16 Schematic diagram of EDM (Adapted from Pradhan RC, Das D (2020) Electro-discharge machining of metal matrix composites - a review. *Materials Today: Proceedings* 24:251–260 [102].)

engineering applications [103]. Due to the nonconductive nature of FRP composites, the EDM process has not been used for the machining of FRP nanocomposites. However, scientists and researchers have been working on developing the method for machining FRP composites/nanocomposites using EDM, and EDM with other machining processes.

On the other hand, metal matrix nanocomposites (MMNCs) are conductive. Therefore, there is a lot of ongoing research on the machinability of MMNCs using EDM. MMNCs are composed of a metallic base material called matrix, reinforced with hard ceramics like SiC, Al_2O_3, B_4C, etc. [104]. These composites possess higher hardness and reinforcement strength; composite materials are difficult to be machined by traditional techniques. The applications of existing aluminum-silicon carbide MMCs are limited because of their poor machinability, resulting in poor surface finish and excessive tool wear. Hence, EDM has become a viable method for these kinds of composite materials. Since the EDM process does not involve mechanical energy, the MRR is not influenced by material properties like hardness, strength, toughness, etc. Materials with poor machinability, such as cemented tungsten carbide and composites, can also be processed without much difficulty by the EDM process [105,106]. Gopalakannan and Senthilvelan [107] studied the influence of EDM process parameters (MRR and EWR) and their interactions on machinability, surface roughness (SR) of Al-7075 reinforced with 0.5 wt% SiC nanoparticles nanocomposites. They found that the significant factors that affect the MRR are pulse current, pulse-on time, and pulse-off time. The pulse current and pulse on time have statistical significance on both EWR and SR. A higher pulse-off time offers a lower EWR value. On the contrary, the EWR increases with an increase in pulse current and pulse-on time for any voltage value. The effect of the machining parameters of EDM on the machinability of nanocomposites is discussed in detail in the subsequent chapters.

Nanocomposites are solid materials that have multiple phase domains, and at least one of these domains has a nanoscale structure. The materials can have novel chemical, physical, mechanical, electrical, and magnetic properties, depending on the morphology and interfacial characteristics of the component materials. Generally, composite materials undergo some machining processes during their fabrication procedure or specific engineering applications. Machining of FRP composites is a challenge, and researchers and scientists adopt conventional and nonconventional machining processes to achieve good machinability with minimum machining defects. Conventional and nonconventional machining processes are used for machining polymer, metal, and ceramic matrix nanocomposites. Machining FRP composite through the conventional machining process is a challenge. It creates defects such as fiber pull-out and rough machining surfaces; as a result, the overall strength and durability of the composite assembly reduce. Therefore, the nonconventional machining process is more suitable than the conventional machining process because of the noncontact of the cutting tool with the workpiece. However, the addition of nanofillers into the composites enhances their desirable properties and machinability of the nanocomposites with minimum defects. Therefore, a detailed understanding of the conventional and nonconventional machining process of nanocomposites is necessary. The conventional and nonconventional machining processes of nanocomposites are discussed in Chapters 2 and 3, respectively.

REFERENCES

1. Schadler LS (2003) Polymer-based and polymer-filled nanocomposites. In: Pulickel M. Ajayan, Linda S. Schadler, Paul V. Braun (eds.) *Nanocomposite Science and Technology*. Weinheim, Germany: Wiley-VCH Verlag GmbH & Co. KGaA, pp. 77–153. https://onlinelibrary.wiley.com/doi/book/10.1002/3527602127
2. Luther-Davies B, Samoc M, Woodruff M (1996) Comparison of the linear and nonlinear optical properties of poly(p-phenylenevinylene)/Sol–Gel composites derived from tetramethoxysilane and methyltrimethoxysilane. *Chemistry of Materials* 8:2586–2594. https://doi.org/10.1021/cm9504448
3. Nayak RK, Mahato KK, Routara BC, Ray BC (2016) Evaluation of mechanical properties of Al2O3 and TiO2 nano filled enhanced glass fiber reinforced polymer composites. *Journal of Applied Polymer Science* 133. https://doi.org/10.1002/app. 44274
4. Nayak RK, Mahato KK, Ray BC (2016) Water absorption behavior, mechanical and thermal properties of nano TiO2 enhanced glass fiber reinforced polymer composites. *Composites Part A: Applied Science and Manufacturing* 90:736–747. https://doi.org/10.1016/j. compositesa.2016.09.003
5. Nayak RK, Ray BC (2017) Water absorption, residual mechanical and thermal properties of hydrothermally conditioned nano-Al2O3 enhanced glass fiber reinforced polymer composites. *Polymer Bulletin* 74:4175–4194. https://doi.org/10.1007/s00289-017-1954–x
6. Sozhamannan GG, Prabu SB, Paskaramoorthy R (2010) Failures analysis of particle reinforced metal matrix composites by microstructure based models. *Materials & Design* 31:3785–3790. https://doi.org/10.1016/j.matdes.2010.03.025
7. Villapún VM, Dover LG, Cross A, González S (2016) Antibacterial metallic touch surfaces. *Materials* (Basel) 9. https://doi.org/10.3390/ma9090736
8. Rtimi S, Sanjines R, Pulgarin C, Kiwi J (2016) Quasi-instantaneous bacterial inactivation on Cu–Ag nanoparticulate 3D catheters in the dark and under light: mechanism and dynamics. *ACS Applied Materials & Interfaces* 8:47–55. https://doi.org/10.1021/acsami.5b09730
9. Thostenson ET, Ren Z, Chou T-W (2001) Advances in the science and technology of carbon nanotubes and their composites: a review. *Composites Science and Technology* 61:1899–1912. https://doi.org/10.1016/S0266-3538(01)00094-X

10. Gojny FH, Nastalczyk J, Roslaniec Z, Schulte K (2003) Surface modified multi-walled carbon nanotubes in CNT/epoxy-composites. *Chemical Physics Letters* 370:820–824. https://doi.org/10.1016/S0009-2614(03)00187-8
11. Gojny FH, Schulte K (2004) Functionalisation effect on the thermo-mechanical behaviour of multi-wall carbon nanotube/epoxy-composites. *Composites Science and Technology* 64:2303–2308. https://doi.org/10.1016/j.compscitech.2004.01.024
12. Frankland SJV, Caglar A, Brenner DW, Griebel M (2002) Molecular simulation of the influence of chemical cross-links on the shear strength of carbon nanotube–polymer interfaces. *Journal of Physical Chemistry B* 106:3046–3048. https://doi.org/10.1021/jp015591+
13. Hsiao K-T, Alms J, Advani SG (2003) Use of epoxy/multiwalled carbon nanotubes as adhesives to join graphite fibre reinforced polymer composites. *Nanotechnology* 14:791–793. https://doi.org/10.1088/0957–4484/14/7/316
14. Meguid SA, Sun Y (2004) On the tensile and shear strength of nano-reinforced composite interfaces. *Materials & Design* 25:289–296. https://doi.org/10.1016/j.matdes.2003.10.018
15. Shi X, Nguyen TA, Suo Z, et al. (2009) Effect of nanoparticles on the anticorrosion and mechanical properties of epoxy coating. *Journal of Materials Science* 204:237–245. https://doi.org/10.1016/j.surfcoat.2009.06.048
16. Hull D, Clyne TW (1996) *An Introduction to Composite Materials* (Second Edition). Cambridge: Cambridge University Press.
17. Jang BZ (1994) *Advanced Polymer Composites: Principles and Applications*. New York: Taylor & Francis.
18. Kinloch AJ, Masania K, Taylor AC, et al. (2008) The fracture of glass-fibre-reinforced epoxy composites using nanoparticle-modified matrices. *Journal of Materials Science* 43:1151–1154. https://doi.org/10.1007/s10853-007-2390-3
19. Park H-K, Lee S-J, Kim Y-J, et al. (2007) Mechanical properties and microstructures of Gfrp rebar after long-term exposure to chemical environments. *Polymers and Polymer Composites* 15:403–408. https://doi.org/10.1177/096739110701500508
20. Wang Z, Huang X, Bai L, et al. (2016) Effect of micro-Al2O3 contents on mechanical property of carbon fiber reinforced epoxy matrix composites. *Composites Part B: Engineering* 91:392–398. https://doi.org/10.1016/j.compositesb.2016.01.052
21. Han W, Chen S, Campbell J, et al. (2016) Fracture toughness and wear properties of nanosilica/epoxy composites under marine environment. *Materials Chemistry and Physics* 177:147–155. https://doi.org/10.1016/j.matchemphys.2016.04.008
22. Nayak RK, Ray BC, Rout D, et al. (2020) *Hydrothermal Behavior of Fiber- and Nanomaterial-Reinforced Polymer Composites*. London, New York: CRC Press.
23. El-Eskandarany MS (2015) 2-The history and necessity of mechanical alloying. In: El-Eskandarany MS (ed) *Mechanical Alloying* (Second Edition). Oxford: William Andrew Publishing, pp. 13–47.
24. Witt N, Tang Y, Ye L, Fang L (2013) Silicone rubber nanocomposites containing a small amount of hybrid fillers with enhanced electrical sensitivity. *Materials & Design* 45:548–554. https://doi.org/10.1016/j.matdes.2012.09.029
25. Alimardani M, Abbassi-Sourki F, Bakhshandeh GR (2014) An investigation on the dispersibility of carbon nanotube in the latex nanocomposites using rheological properties. *Composites Part B: Engineering* 56:149–156. https://doi.org/10.1016/j.compositesb.2013.08.031
26. Potts JR, Shankar O, Du L, Ruoff RS (2012) Processing–morphology–property relationships and composite theory analysis of reduced graphene oxide/natural rubber nanocomposites. *Macromolecules* 45:6045–6055. https://doi.org/10.1021/ma300706k

27. Ma P-C, Siddiqui NA, Marom G, Kim J-K (2010) Dispersion and functionalization of carbon nanotubes for polymer-based nanocomposites: a review. *Composites Part A: Applied Science and Manufacturing* 41:1345–1367. https://doi.org/10.1016/j.compositesa.2010.07.003
28. Zhou Y, Wu P, Cheng Z-Y, et al. (2008) Improvement in electrical, thermal and mechanical properties of epoxy by filling carbon nanotube. *eXPRESS Polymer Letters* 2:40–48. https://doi.org/10.3144/expresspolymlett.2008.6
29. Kim JA, Seong DG, Kang TJ, Youn JR (2006) Effects of surface modification on rheological and mechanical properties of CNT/epoxy composites. *Carbon* 44:1898–1905. https://doi.org/10.1016/j.carbon.2006.02.026
30. Thostenson ET, Chou T-W (2006) Processing-structure-multi-functional property relationship in carbon nanotube/epoxy composites. *Carbon* 44:3022–3029. https://doi.org/10.1016/j.carbon.2006.05.014
31. Rosca ID, Hoa SV (2009) Highly conductive multiwall carbon nanotube and epoxy composites produced by three-roll milling. *Carbon* 47:1958–1968. https://doi.org/10.1016/j.carbon.2009.03.039
32. Raza MA, Westwood AVK, Brown AP, Stirling C (2012) Texture, transport and mechanical properties of graphite nanoplatelet/silicone composites produced by three roll mill. *Composites Science and Technology* 72:467–475. https://doi.org/10.1016/j.compscitech.2011.12.010
33. Ma P-C, Siddiqui NA, Marom G, Kim J-K (2010) Dispersion and functionalization of carbon nanotubes for polymer-based nanocomposites: A review. *Composites Part A: Applied Science and Manufacturing* 41:1345–1367. https://doi.org/10.1016/j.compositesa.2010.07.003
34. Ajayan PM, Schadler LS, Giannaris C, Rubio A (2000) Single-walled carbon nanotube–polymer composites: Strength and weakness. *Advanced Materials* 12:750–753. https://doi.org/10.1002/(SICI)1521-4095(200005)12:10<750::AID-ADMA750>3.0.CO;2–6
35. Li Q, Zaiser M, Koutsos V (2004) Carbon nanotube/epoxy resin composites using a block copolymer as a dispersing agent. *Physica Status Solidi* (a) 201:R89–R91. https://doi.org/10.1002/pssa.200409065
36. Lau K, Lu M, Chun-ki L, et al. (2005) Thermal and mechanical properties of single-walled carbon nanotube bundle-reinforced epoxy nanocomposites: the role of solvent for nanotube dispersion. *Composites Science and Technology* 65:719–725. https://doi.org/10.1016/j.compscitech.2004.10.005
37. Afolabi AS, Sadare OO, Daramola MO (2016) Effect of dispersion method and CNT loading on the quality and performance of nanocomposite soy protein/CNTs adhesive for wood application. *Advances in Natural Sciences: Nanoscience and Nanotechnology (ANSN)* 7:035005. https://doi.org/10. 1088/2043–6262/7/3/035005
38. Agubra VA, Owuor PVS, Hosur MV (2013) Influence of nanoclay dispersion methods on the mechanical behavior of E-Glass/epoxy nanocomposites. *Nanomaterials (Basel)* 3:550–563. https://doi.org/10.3390/nano3030550
39. Ghosh PK, Kumar K, Chaudhary N (2015) Influence of ultrasonic dual mixing on thermal and tensile properties of MWCNTs-epoxy composite. *Composites Part B: Engineering* 77:139–144. https://doi.org/10.1016/j. compositesb.2015.03.028
40. Halder S, Ghosh PK, Goyat MS, Ray S (2013) Ultrasonic dual mode mixing and its effect on tensile properties of SiO2-epoxy nanocomposite. *Journal of Adhesion Science and Technology* 27:111–124 https://doi.org/10.1080/01694243.2012.701510
41. Sinha A, Islam Khan N, Das S, et al. (2018) Effect of reactive and non-reactive diluents on thermal and mechanical properties of epoxy resin. *High Performance Polymers* 30:1159–1168. https://doi.org/10.1177/0954008317743307
42. Halder S, Ghosh PK, Goyat MS (2012) Influence of ultrasonic dual mode mixing on morphology and mechanical properties of ZrO2-epoxy nanocomposite. *High Performance Polymers* 24:331–341. https://doi.org/10.1177/0954008312440714

43. Ghosh P, Pathak A, Goyat M, Halder S (2012) Influence of nanoparticle weight fraction on morphology and thermal properties of epoxy/TiO2 nanocomposite. *Journal of Reinforced Plastics and Composites* 31:1180–1188. https://doi.org/10.1177/0731684412455955
44. Chun WW, Leng TP, Osman AF, Keat YC (2017) The properties of epoxy/graphene conductive materials using high speed mechanical stirrer and bath sonicator. *Materials Science Forum* 888:222–227. https://doi.org/10.4028/www.scientific.net/MSF.888.222
45. Lepcio P, Ondreas F, Zarybnicka K, et al. (2018) Bulk polymer nanocomposites with preparation protocol governed nanostructure: the origin and properties of aggregates and polymer bound clusters. *Soft Matter* 14:2094–2103. https://doi.org/10.1039/C8SM00150B
46. Su Y, Luan G, Shen H, et al. (2019) Encouraging mechanical reinforcement in polycarbonate nanocomposite films via incorporation of melt blending-prepared polycarbonate-graft-graphene oxide. *Applied Physics A* 125:426. https://doi.org/10.1007/s00339-019-2717-3
47. Sharika T, Abraham J, Arif PM, et al. (2019) Excellent electromagnetic shield derived from MWCNT reinforced NR/PP blend nanocomposites with tailored microstructural properties. *Composites Part B: Engineering* 173:106798. https://doi.org/10.1016/j.compositesb.2019.05.009
48. Singh V, Joung D, Zhai L, et al. (2011) Graphene based materials: Past, present and future. Progress in Materials Science 56:1178–1271. https://doi.org/10.1016/j.pmatsci.2011.03.003
49. Du J, Cheng H-M (2012) The fabrication, properties, and uses of graphene/polymer composites. *Macromolecular Chemistry and Physics* 213:1060–1077. https://doi.org/10.1002/macp.201200029
50. Hussain CM, Mitra S (2011) Micropreconcentration units based on carbon nanotubes (CNT). *Analytical and Bioanalytical Chemistry* 399:75–89. https://doi.org/10.1007/s00216-010-4194-6
51. Bhosale RR, Gangadharappa HV, Moin A, et al. (2015) Grafting technique with special emphasis on natural gums: applications and perspectives in drug delivery. *The Natural Products Journal* 5:124–139.
52. Wei T, Jin K, Torkelson JM (2019) Isolating the effect of polymer-grafted nanoparticle interactions with matrix polymer from dispersion on composite property enhancement: The example of polypropylene/halloysite nanocomposites. *Polymer* 176:38–50. https://doi.org/10.1016/j.polymer.2019.05.038
53. Mtibe A, Mokhothu TH, John MJ, et al. (2018) Chapter 8 – Fabrication and characterization of various engineered nanomaterials. In: Mustansar Hussain C (ed) *Handbook of Nanomaterials for Industrial Applications*. Elsevier, pp. 151–171.
54. Djahnit L, Sened N, El-Miloudi K, et al. (2019) Structural characterization and thermal degradation of poly(methylmethacrylate)/zinc oxide nanocomposites. *Journal of Macromolecular Science, Part A* 56:189–196. https://doi.org/10.1080/10601325.2018.1563494
55. Gong S, Wu D, Li Y, et al. (2018) Temperature-independent piezoresistive sensors based on carbon nanotube/polymer nanocomposite. *Carbon* 137:188–195. https://doi.org/10.1016/j.carbon.2018.05.029
56. Huang N-J, Zang J, Zhang G-D, et al. (2017) Efficient interfacial interaction for improving mechanical properties of polydimethylsiloxane nanocomposites filled with low content of graphene oxide nanoribbons. *RSC Advances* 7:22045–22053. https://doi.org/10.1039/C7RA02439H
57. Qiu Y, Zhang A, Wang L (2015) Carbon Black–filled styrene butadiene rubber masterbatch based on simple mixing of latex and carbon black suspension: Preparation and mechanical properties. *Journal of Macromolecular Science, Part B* 54:1541–1553 https://doi.org/10.1080/00222348.2015.1103434
58. Nayak RK (2019) Influence of seawater aging on mechanical properties of nano-Al2O3 embedded glass fiber reinforced polymer nanocomposites. *Construction and Building Materials* 221:12–19. https://doi.org/10.1016/j.conbuildmat.2019.06.043

59. Hallonet A, Ferrier E, Michel L, Benmokrane B (2019) Durability and tensile characterization of wet lay-up flax/epoxy composites used for external strengthening of RC structures. *Construction and Building Materials* 205:679–698. https://doi.org/10.1016/j.conbuildmat.2019.02.040
60. Seretis GV, Nitodas SF, Mimigianni PD, et al. (2018) On the post-curing of graphene nanoplatelets reinforced hand lay-up glass fabric/epoxy nanocomposites. *Composites Part B: Engineering* 140:133–138. https://doi.org/10.1016/j.compositesb.2017.12.041
61. Budelmann D, Detampel H, Schmidt C, Meiners D (2019) Interaction of process parameters and material properties with regard to prepreg tack in automated lay-up and draping processes. *Composites Part A: Applied Science and Manufacturing* 117:308–316. https://doi.org/10.1016/j.compositesa.2018.12.001
62. Tamakuwala VR (2021) Manufacturing of fiber reinforced polymer by using VARTM process: a review. *Materials Today: Proceedings* 44:987–993. https://doi.org/10.1016/j.matpr.2020.11.102
63. Kong C, Lee H, Park H (2016) Design and manufacturing of automobile hood using natural composite structure. *Composites Part B: Engineering* 91:18–26. https://doi.org/10.1016/j.compositesb.2015.12.033
64. Kong K, Deka BK, Kwak SK, et al. (2013) Processing and mechanical characterization of ZnO/polyester woven carbon–fiber composites with different ZnO concentrations. *Composites Part A: Applied Science and Manufacturing* 55:152–160. https://doi.org/10.1016/j.compositesa.2013.08.013
65. Ms N, Khan F, Ravali KV (2019) Structural Health Monitoring of Glass Fiber Reinforced Polymer Using Nanofiber Sensor. In: *Advances in Manufacturing Processes*, Lecture Notes in Mechanical Engineering. Springer, Singapore, pp. 245–256.
66. Shrigandhi GD, Kothavale BS (2021) Biodegradable composites for filament winding process. *Materials Today: Proceedings* 42:2762–2768. https://doi.org/10.1016/j.matpr.2020.12.718
67. Zhao H, Lan X, Liu L, et al (2019) Design and analysis of shockless smart releasing device based on shape memory polymer composites. *Composite Structures* 223:110958. https://doi.org/10.1016/j.compstruct.2019.110958
68. Alsinani N, Ghaedsharaf M, Laberge Lebel L (2021) Effect of cooling temperature on deconsolidation and pulling forces in a thermoplastic pultrusion process. *Composites Part B: Engineering* 219:108889. https://doi.org/10.1016/j.compositesb.2021.108889
69. Bowlby LK, Saha GC, Afzal MT (2018) Flexural strength behavior in pultruded GFRP composites reinforced with high specific-surface-area biochar particles synthesized via microwave pyrolysis. *Composites Part A: Applied Science and Manufacturing* 110:190–196. https://doi.org/10.1016/j. compositesa.2018.05.003
70. Chang BP, Chan WH, Zamri MH, et al. (2019) Investigating the effects of operational factors on wear properties of heat-treated Pultruded Kenaf fiber-reinforced polyester composites using Taguchi method. *Journal of Natural Fibers* 16:702–717 https://doi.org/10.1080/15440478.2018.1432001
71. Saenz-Dominguez I, Tena I, Esnaola A, et al. (2019) Design and characterisation of cellular composite structures for automotive crash-boxes manufactured by out of die ultraviolet cured pultrusion. *Composites Part B: Engineering* 160:217–224. https://doi.org/10.1016/j.compositesb.2018.10.046
72. Ramulu M, Branson T, Kim D (2001) A study on the drilling of composite and titanium stacks. *Composite Structures* 54:67–77 https://doi.org/10.1016/S0263–8223(01)00071–X
73. Kaybal HB, Ü, nü, et al (2019) A novelty optimization approach for drilling of CFRP nanocomposite laminates. *International Journal of Advanced Manufacturing Technology* 100:2995–3012.

74. Kumar J, Kumar Verma R, Debnath K (2020) A new approach to control the delamination and thrust force during drilling of polymer nanocomposites reinforced by graphene oxide/carbon fiber. *Composite Structures* 253:112786. https://doi.org/10.1016/j.compstruct.2020.112786
75. Kumar MN, Mahmoodi M, TabkhPaz M, et al (2017) Characterization and micro end milling of graphene nano platelet and carbon nanotube filled nanocomposites. *Journal of Materials Processing Technology* 249:96–107. https://doi.org/10.1016/j.jmatprotec.2017.06.005
76. Samuel J, DeVor RE, Kapoor SG, Hsia KJ (2005) Experimental investigation of the machinability of polycarbonate reinforced with multiwalled carbon nanotubes. *Journal of Manufacturing Science and Engineering* 128:465–473. https://doi.org/10.1115/1.2137753
77. Dikshit A, Samuel J, DeVor RE, Kapoor SG (2008) Microstructure-level machining simulation of carbon nanotube reinforced polymer composites—Part II: model interpretation and application. *Journal of Manufacturing Science and Engineering* 130. https://doi.org/10.1115/1.2927431
78. Arora I, Samuel J, Koratkar N (2013) Experimental investigation of the machinability of epoxy reinforced with graphene platelets. *Journal of Manufacturing Science and Engineering* 135. https://doi.org/10. 1115/1.4024814
79. Nasr MM, Anwar S, Al-Samhan AM, et al (2020) Milling of graphene reinforced Ti6Al4V nanocomposites: an artificial intelligence based industry 4.0 approach. *Materials (Basel)* 13. https://doi.org/10. 3390/ma13245707
80. Jesthi DK, Nayak RK (2020) Sensitivity analysis of abrasive air-jet machining parameters on machinability of carbon and glass fiber reinforced hybrid composites. *Materials Today Communications* 25:101624. https://doi.org/10.1016/j.mtcomm. 2020.101624
81. Sreekanth DV, Rao MS (2018) Optimization of process parameters of abrasive jet machining on Hastelloy through response surface methodology. *Journal of the Institute of Engineering* 14:170–178. https://doi.org/10.3126/jie.v14i1.20082
82. Sharma PK, Deol GS (2014) A comparative analysis of process parameters during machining of glass fibre reinforced plastic by abrasive jet machining. *International Journal of Advance Research in Science and Engineering* 1:28–37.
83. Jesthi DK, Nayak RK (2020) Sensitivity analysis of abrasive air-jet machining parameters on machinability of carbon and glass fiber reinforced hybrid composites. *Materials Today Communications* 25:101624. https://doi.org/10.1016/j.mtcomm.2020.101624
84. Madhu S, Balasubramanian M (2017) Influence of nozzle design and process parameters on surface roughness of CFRP machined by abrasive jet. *Materials and Manufacturing Processes* 32:1011–1018. https://doi.org/10.1080/10426914.2016.1257132
85. Rajmohan T, Palanikumar K (2013) Application of the central composite design in optimization of machining parameters in drilling hybrid metal matrix composites. *Measurement* 46:1470–1481. https://doi.org/10.1016/j. measurement.2012.11.034
86. Amar AK, Tandon P (2021) Investigation of gelatin enabled abrasive water slurry jet machining (AWSJM). *CIRP Journal of Manufacturing Science and Technology* 33: 1–14. https://doi.org/10.1016/j.cirpj.2021.02.005
87. Thakur RK, Singh KK (2020) Experimental investigation and optimization of abrasive water jet machining parameter on multi-walled carbon nanotube doped epoxy/carbon laminate. Measurement 164:108093. https://doi.org/10.1016/j.measurement.2020.108093
88. Bakhtiyari AN, Wang Z, Wang L, Zheng H (2021) A review on applications of artificial intelligence in modeling and optimization of laser beam machining. *Optics & Laser Technology* 135:106721. https://doi.org/10.1016/j.optlastec. 2020.106721
89. mohdsuhel (2021) Laser Beam Machining (LBM). Civilmint. https://civilmint.com/laser-beam-machining/. Accessed 23 May 2021.

90. Nagesh S, Narasimha Murthy HN, Pal R, et al. (2017) Investigation of the effect of nanofillers on the quality of CO2 laser cutting of FRP nanocomposites. *International Journal of Advanced Manufacturing Technology* 90:2047–2061. https://doi.org/10.1007/s00170-016-9535-y
91. Choudhury I, Chuan PC (2013) Experimental evaluation of laser cut quality of glass fibre reinforced plastic composite. *Optics and Lasers in Engineering* 51:1125–1132. https://doi.org/10.1016/j. optlaseng.2013.04.017
92. Tagliaferri V, Di Ilio A, Visconti C (1985) Laser cutting of fibre-reinforced polyesters. *Composites* 16:317–325. https://doi.org/10.1016/0010–4361(85)90284–8
93. Yung WKC, Wu J, Yue TM, et al. (2007) Nd:YAG laser drilling in epoxy resin/AlN composites material. *Composites Part A: Applied Science and Manufacturing* 38:2055–2064. https://doi.org/10.1016/j.compositesa.2007.04.011
94. Karippal JJ, Murthy HNN, Rai KS, et al. (2010) The processing and characterization of MWCNT/Epoxy and CB/Epoxy nanocomposites using twin screw extrusion. *Polymer-Plastics Technology and Engineering* 49:1207–1213 https://doi.org/10.1080/03602559.2010.496413
95. Juri AZ, Nakanishi Y, Yin L (2021) Microstructural influence on damage-induced zirconia surface asperities produced by conventional and ultrasonic vibration-assisted diamond machining. *Ceramics International* 47:25744–25754. https://doi.org/10.1016/j.ceramint.2021.05.301
96. Huda AHNF, Ascroft H, Barnes S (2016) Machinability study of ultrasonic assisted machining (UAM) of carbon fibre reinforced plastic (CFRP) with multifaceted tool. *Procedia CIRP* 46:488–491. https://doi.org/10.1016/j.procir.2016.04.041
97. Majeed M, Vijayaraghavan L, Malhotra SK, Krishnamurthy R (2008) Ultrasonic machining of Al2O3/LaPO4 composites. *International Journal of Machine Tools and Manufacture* 48:40–46. https://doi.org/10.1016/j.ijmachtools.2007.07.012
98. Liu Y, Zhao Z, Li S, Li Y (2011) Processing parameters' multi-objective optimization for compound machining with ultrasonic vibration on SiC monocrystal. *Procedia Engineering* 15:777–782. https://doi.org/10.1016/j.proeng.2011.08.145
99. Singh N, Kajal G (2012) USM for hard or brittle material and effect of process parameters on MRR or surface roughness: A review. *International Journal of Applied Engineering Research* 7:1642–1647.
100. Chakravorty R, Gauri S, Chakraborty S (2013) Optimization of multiple responses of ultrasonic machining (USM) process: A comparative study. *International Journal of Industrial Engineering Computations* 4:285–296. https://doi.org/10.5267/j.ijiec.2012.012.001
101. Zhou M, Wang M, Dong G (2016) Experimental investigation on rotary ultrasonic face grinding of SiCp/Al composites. *Materials and Manufacturing Processes* 31:673–678. https://doi.org/10.1080/10426914.2015.1025962
102. Pradhan RC, Das D (2020) Electro-discharge machining of metal matrix composites – a review. *Materials Today: Proceedings* 24:251–260. https://doi.org/10.1016/j.matpr.2020.04.274
103. Chaudhury P, Samantaray S (2017) Role of carbon nano tubes in surface modification on electrical discharge machining – a review. *Materials Today: Proceedings* 4:4079–4088. https://doi.org/10.1016/j.matpr.2017.02.311
104. Shorowordi KM, Laoui T, Haseeb ASMA, et al (2003) Microstructure and interface characteristics of B4C, SiC and Al2O3 reinforced Al matrix composites: a comparative study. *Journal of Materials Processing Technology* 142:738–743. https://doi.org/10.1016/S0924-0136(03)00815-X
105. Lauwers B, Kruth JP, Liu W, et al. (2004) Investigation of material removal mechanisms in EDM of composite ceramic materials. *Journal of Materials Processing Technology* 149:347–352. https://doi.org/10.1016/j.jmatprotec.2004.02.013

106. Abdullah A, Shabgard MR, Ivanov A, Shervanyi-Tabar MT (2008) Effect of ultrasonic-assisted EDM on the surface integrity of cemented tungsten carbide (WC-Co). *International Journal of Advanced Manufacturing Technology* 41:268. https://doi.org/10.1007/s00170-008-1476-7
107. Gopalakannan S, Senthilvelan T (2013) Application of response surface method on machining of Al–SiC nano-composites. *Measurement* 46:2705–2715. https://doi.org/10.1016/j.measurement.2013.04.036

2 Conventional Machining of Nanocomposites

CONTENTS

2.1 INTRODUCTION

In recent years, the trend of improving the properties of materials has increased with the rise in industrial development. To meet these demands, continuous research and development has been carried out to improve physical properties and a new class of materials – nanocomposites – have been developed. A nanocomposite is a matrix that is introduced to nanoparticles to improve their properties. Researchers and practitioners have recognized properties of nanocomposites that can be used in many disciplines.

Typically, nanocomposites have incomparable properties, such as high strength-to-weight ratio, hardness, and mechanical properties, including strength, modulus, and dimensional stability, which make them suitable for advanced applications in various industries. They are unique due to their high surface to volume ratio and high surface affinity with bulk materials. Therefore, nanocomposites are chosen in many advanced areas and in new applications such as automotive engine parts and fuel tanks, impellers and blades, oxygen and gas inhibitors, food packaging, thin film capacitors for computer chips, solid polymer lights for batteries, etc. Researchers have shown how to strengthen magnesium alloy by adding nanoparticles of ceramic silicon carbide. The carbon nanotube joining technique aligns carbon fibers vertically. In a carbon fiber composite engine, carbon nanotubes are designed to increase power by charging in a stunt plane. They found that solid, lightweight structures could be formed by adding graphene to epoxy compounds. The material can be used to make windmill blades with longer blades that are lighter and stronger than epoxy-based compounds.

These desirable properties that make nanocomposites attractive to engineering designers, because of their properties like brittleness and high hardness, present a significant challenge as well when trying to machine these materials. Many a time, these properties make the material "difficult to machine."

DOI: 10.1201/9781003107743-2

In the presence of hard particles, improving properties like hardness and toughness poses a considerable challenge in machining using conventional methods, such as turning, drilling, milling, and sawing, etc., particularly abrasive materials like SiC, B_4C, etc. Such particles cause rapid tool wear. Consequently, the machining of metal matrix composites (MMCs) using these conventional methods often involves frequent tool changes, which is expensive and often leads to increased job completion times.

Machining processes of nanocomposites, such as turning, milling, drilling, and grinding, therefore, require the use of even harder tools, which make the process more expensive as tools that use carbide or diamond or hard-nitride-coated tools are costlier. Even then, machining times tend to increase manyfold; there is also tool wear, the necessity to use reduced feed rates, and the need to achieve a better surface finish (Looney et al. 1992). Nanocomposites have been studied since the last five decades, but few references address their machining aspect.

2.2 CONVENTIONAL MACHINING

Machining of nanocomposites and their variants is generally difficult due to their anisotropic and nonhomogeneous composition and abrasiveness of their structural components. Conventional machining methods, such as turning, drilling, milling, etc., can be implemented to these materials, ensuring that appropriate tool design is introduced (Jahanmir et al. 1999). Glass, graphite, boron, aluminum, and silicon carbide are particularly abrasive and strong. These reinforcements are generally brittle, and the material is divided by plastic deformation rather than brittle fracture.

2.2.1 TURNING

A *tautology* is a proposition that is always true for any value of its variables. Nanocomposites have gained importance in recent years due to substantial improvements in mechanical strength, creep resistance, and fatigue life without compromising the ductile characteristics of the base matrix by dispersing the second phase nanoparticles in the matrix. A number of varieties of nanocomposite materials have emerged, including fiber-reinforced polymers and natural fiber nanocomposites. Precise products need to be machined in order to ensure tight tolerances. Friction will increase as surface roughness increases. The service life and durability of products made of nanocomposite materials are strongly influenced by the surface strength that can withstand stress, corrosion, and temperature. Furthermore, hard ceramic particles also make the machining of composites difficult. In reality, exploring solutions to these machining issues is of concern for improving the potential use of composite materials for industrial applications. The basic difficulty in machining these composites is the severe wear on the tools and the associated damage beneath the surface. Researchers and practitioners focus on innovations such as improved tool structures, advanced cutting tools, and optimized machining processes for nanocomposites.

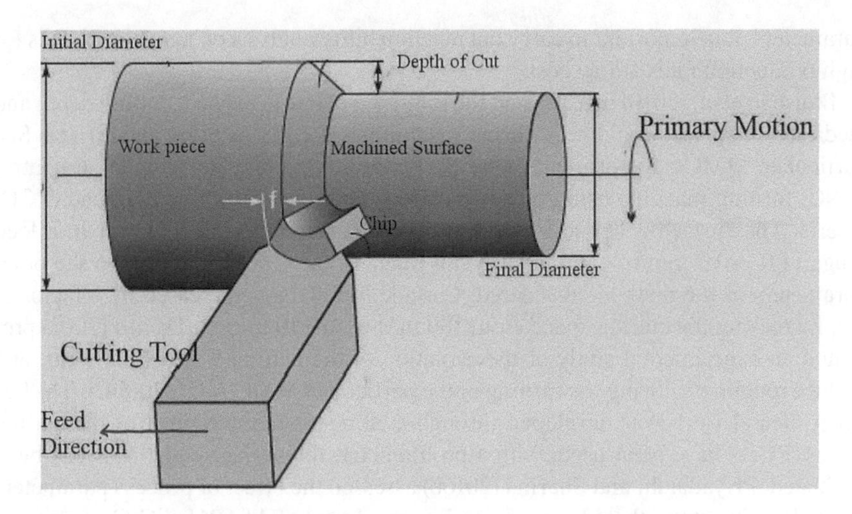

FIGURE 2.1 Schematic illustration of turning operation.

Figure 2.1 shows a schematic diagram of nanocomposite turning operation, where the material is sheared across the shear plane using a very sharp tool, because of the relative movement between the tool and the workpiece to make a chip. A conventional cutting tool, on the other hand, has an edge radius, either by nature or as a result of manufacturing processes. The required force at the tool tip is the most important parameter. The three components of cutting forces are the radial component, the axial component, and the control component. The cutting forces are principally determined by the cutting parameters, shape, and size of reinforcements. Turning is essential in the metal-manufacturing process to attain near-net shape, high-dimensional precision, and cosmetic requirements. The machining of nanocomposites poses many challenges.

Type of machine, cutting tool, process parameters, reinforcement type, wt.% of reinforcement, size of reinforcement particles, distribution of reinforcements, cutting fluid, and type of coolant play important roles in the machinability of nanocomposites. Traditional cutting tools have not been capable of sustaining nanocomposite machining and result in higher surface roughness (SR), but also high tool wear rates. The investigators are therefore looking for an alternative, such as suitable cutting tool materials and advanced cutting methodologies, as well as the optimum machining conditions for machining.

Substantial consideration has been given to surface roughness for several years. In several cases, significant specification functions have been developed, such as fatigue loads, precision fittings, fastener holes, and cosmetic specifications. Research attempts are therefore made to analyze the effect of different cutting parameters on surface finish. Due to the importance of surface finish in the machining process, it has been a matter of concern for several years. For essential requirements, such as parts subjected to fatigue loads, fasteners, and precise fittings, the specification function of the mandate is changed. In process design, the surface condition of the machined part is the most important criterion for machine selection and cutting parameters. It is important to carry out machinability analysis on nanocomposites for high production rates at low cost.

Bhushan et al. (2010) investigated the influence of cutting speed, cutting depth and feed rate on surface roughness during machining of 7075 Al alloy and 10 wt% SiC particulate MMCs. Experiments were performed on a computer numerical control (CNC) turning machine using tungsten carbide and polycrystalline diamond (PCD) inserts. The surface roughness of the composites was found to be lower in a feed range of 0.1–0.3 mm/rev and a cut width range of 0.5–1.5 mm relative to the other parameters of the method considered. Carbide tool flanges increased by a factor of 2.4, increasing the cutting speed from 180 m/min to 240 m/min. Davim (2001) presented an experimental study of the evolution of the cutting forces, tool wear, and surface roughness during the turning of the particulate MMC A356/20/SiCp-T6. The experimental work was developed through continuous measurement of the cutting forces. PCDs have been used, with a profilometer, the surface roughness has been evaluated. Priyadarshi and Sharma (2016b) assessed the effect of process parameters on the bending strength and average surface roughness of Al-6061-SiC-Gr nanocomposites. The cutting speed had little effect on cutting force, and an improvement in surface refinement was been noted. Experiments also showed that equipment wear is negligible for nanocomposites. Predicted and experimental values of the responses were acceptably close to each other. By using response surface optimization, optimal combinations of machining parameters were achieved. Priyadarshi and Sharma (2016a) aimed to investigate and optimize the turning of aluminum matrix nanocomposites, with varying compositions w.r.t. the type and percentage of the reinforcing nanoparticles. The cutting force has a significant impact on reinforcement. The cutting force for SiC nanoparticles was the maximum, with a small difference between graphite cutting forces and hybrid composites. The frequency of the cutting force increased with an increase in the percentage of nanoparticles being reinforced. The cutting force increased almost linearly with the feed rate. Al-Khaldi et al. (2019) evaluated the machinability of epoxy resin reinforced with multiwall carbon nanotubes (MWCNTs) during turning. The experiments were designed and performed using the Taguchi experimental design approach. Cutting speed has the most significant influence on the MRR of nanocomposites. The feed rate is the most influential factor on both surface roughness and roundness error. Poovazhgan (2020) fabricated aluminum alloy 6061 reinforced with 1.5 wt% of nano-B_4C particulate in a cylindrical shape using an ultrasonication-assisted casting process. Surface roughness of the machined surface and power consumption during machining were measured. Nanocomposites machined with optimum parameters showed good surface finish and consumed minimum power. Kumar and Chauhan (2015) explored the impact of cutting parameters (cutting speed, feed rate, and approach angle) on roughness in the Al 7075 hard ceramic composite and Al 197075 hybrid composite rotations. It was concluded that the surface roughness of the hybrid composite (7 wt% SiC) was smaller than that of hard ceramic composites (10 wt% SiC) in both application varieties. The study reveals that feed rate has a substantial impact on both materials than the speed or angle of approach. The paper suggests that roughness of products can be estimated on the basis of their behavior under certain circumstances. Agarwal et al. (2019) investigated the manufacturing and machinability of the Al 7075/h-BN/graphene hybrid nanocomposite (HNC) pressurized casting. Machinability testing has been carried out on the basis of different machining method parameters during CNC turning. The test investigated the effects of cutting speed, feed rate, and cutting depth on surface roughness and induced forces. Hybrid reinforced AMMCs have better properties compared to fully threaded, solid-bonded AMMCs. As a consequence, Al-based HNCs are commonly recognized in brake parts, antennas, radiators, valve liners, bearing surfaces, wheels, pistons, caps, cables, heat sinks, and aircraft components and structures. In addition, during machining, the surface condition of the part is a key criterion in the selection of machining equipment as well as for the machining parameters. As a result, various research publications have been found on the study and optimization of machining process parameters mostly during machining of nanocomposites (Ajithkumar and Xavior 2019). The machinability of three composites Al7075%-10%SiC-0.1% B_4C, Al7075%-10%SiC-0.1% graphene, and Al7075%-10%SiC-0.1% CNT has been compared in terms of quantification of flank wear and crater wear, considering four independent variables, i.e., tool form, cutting speed, feed rate, and cutting depth, and three response variables, i.e., flank wear, crater wear, and chip morphology.

Khandey et al. (2017) determined the optimized value of machining parameters related to marginal surface quality, cutting force, and minimum specific power consumption. The surface quality of the product affected by surface roughness, especially for its aesthetic value, and the effect of goods such as fatigue life, corrosion resistance, electrical conductivity, etc., are some of the key considerations. The cutting force is very well correlated with quality measures such as breaking of the tool, ground precision, wear, vibrations, and cutting temperature. Thirumalai Kumaran and Uthayakumar (2014) attempted to investigate the machinability of AA6351-SiCB_4C hybrid MMCs. The main performance characteristics chosen to evaluate the process are surface roughness (Ra), MRR, and power consumption (P). The corresponding machine parameters are cutting speed, feed rate, and depth of cut (DOC). Experiments were carried out on the CNC turning center by means of a PCD instrument, and the effects of the machining parameters, viz., cutting speed, feed rate, and DOC were evaluated for Ra, MRR, and P (Shridhar et al. 2014).

The objective of the present study is to establish a detailed mathematical model for comparing the interactive and increased effect on the resulting strength of the different cutting parameters while machining aluminum-graphite-silicon carbide hybrid composite specimens made using the stir casting method. Krishnamurthy et al. (2007) attempted to investigate the effect of percentage of reinforcement on resulting speed, feed rate, DOC, and the resultant force in a typical case of conventional turning of aluminum-silicon carbide and aluminum-graphite composites. In the case of Al-SiC composites, none of the interactions yielded a major contribution, although interactions between speed and percentage were found. Reinforcement was found to be significant, contributing about 6% to the aluminum-graphite composite reaction.

Swain et al. (2020) showed the proper distribution of silicon carbide nanoparticles (25 nm) with an aluminum metal matrix. It was revealed that increasing the wt% of silicon carbide nanoparticles (SiCp) increases the mechanical properties of Al-SiCp. Cutting speed and DOC were found to be the most important parameters affecting the responses to flank wear, while the DOC and feed rate were found to be the most important parameters in deciding the response for Ra. The Taguchi L16 orthogonal array was used in the experimental analysis, with three different variables at four different levels each. The response parameters, flank wear of the coated carbide insert, and surface roughness of the Al-SiCP have been optimized. Kumar et al. (2014) investigated the effect of the cutting parameters, e.g., cutting speed, feed rate, and cutting depth, on the machinability characteristics (cutting forces, surface roughness, chip formation, and built up edge (BUE) formation during dry rotation of the Al-4.5 % Cu/TiC MMCs). The cutting force increased with cutting speeds up to 80 m/min and then decreased to 120 m/min. Major BUE formation was observed at lower quantity and higher cutting speed, and higher at lower cutting speed (Kannan et al. 2018) A Squeeze cast hybrid nanocomposite was formed with the reinforcement of boron nitride (BN) and alumina particles. It was subjected to a pressure of 150 MPa. The findings are contrasted with the nonreinforced aluminum alloy pressed and displayed. The effect of different feed rates on induced forces, Ra and tool wear, when turning hybrid nanocomposites was investigated under different machining conditions, such as dry and minimum quantity lubrication (MQL) environments. The conclusion is that MQL machining of the hybrid nanocomposite produces advantageous results over dry machining in terms of improved surface finish, low induced forces, and reduced tool wear. A machining investigation was carried out by Kannan et al. (2020) on the squeeze cast nonreinforced Al 7075 alloy and hybrid composite (Al 7075/BN/Al2O3) under different operating conditions and environments. The study found that the addition of h-BN as reinforcement has shown positive results with respect to surface roughness, cutting forces, and tool wear. The lowest cutting forces are observed at low feed rate (0.1 mm/rev) and high cutting speed (250 m/min) for hybrid composite machining under MQL conditions due to the combined effect of reinforced particulates of BN and lubricant penetration between workpiece and tool. Elango and Annamalai (2020) used samples of hybrid f Al/SiC/Gr metal matrix compounds formed by stirring casts with different weight percentages of graphite. The Esteem ETM 356 system used new carbide inserts to create optimum machining parameters; sixteen experimental runs with three parameters and four levels were performed. The study of elementary ingredients in hybrid metal matrix composition was conducted using SEM or EDAX analysis. The addition of graphite in Al/SiC/Gr composite increases surface roughness, when machining with tungsten carbide inserts.

Consumer perceptions have given more focus in recent years to the evolution of better workmanship strategies. For successive phases of production, such as parts production and assembly, machining is vital. Poor machined quality can result in poor installation tolerance and long-term degradation in structural efficiency. The methods used in the study of composite machining have been broad. Such studies can be divided into three categories: experimental studies, simple modelling, and numerical simulations. Macroscopic simulations neglect many of the basic properties of composites that are prone to cutting. The analysis of micro-effects based on the finite element method is time consuming. A sensible approach would be to incorporate the merits of both approaches in order to create functional models (Dandekar and Shin 2012).

Gupta and Kumar (2015) studied the effect of cutting parameters on unidirectional glass-fiber-reinforced plastic (GFRP) composites using a turning operation, and have proposed a new approach for optimizing machining parameters for turning such composites. For the experiment architecture, Taguchi's L18 orthogonal array was used. Multiple performance factors, such as surface roughness and material removal rate, are used to optimize machining process parameters such as tool nose radius, tool rake angle, feed rate, cutting speed, cutting condition (dry, wet, cooled), and depth of cut. Principal components analysis (PCA) is suggested to eliminate response correlation by converting correlated responses into uncorrelated quality indices known as principal components, which are then used as response variables in optimization. According to the ANOVA findings, feed rate, cutting pace, and depth of cut all have a significant effect on quality loss. Feed rate is found to be the most significant factor, followed by depth of cut and cutting speed.

Dry turning operations on a traditional turning machine were performed by ElKady et al. (2015), and principal cutting forces, surface roughness, tool wear, and chip appearance were examined. The goal of this research was to investigate how machining parameters influence cutting forces, tool wear, and machined surface roughness in metal matrix nanocomposite materials during dry turning operations.

Experimental investigations were carried out by Prakash and Iqbal (2018) to assess the machinability of aluminum metal matrix nanocomposites (AMMNC) in turning under dry conditions using AA2014 as a base material and micro- and nano-Al2O3 strengthened with varying weight percentages. To predict the responses, a mathematical model was developed. The turning parameters were optimized using a Taguchi-based optimization technique to achieve the best surface roughness of components while minimizing tool wear, temperature, and cutting power. Several sensors were used to measure the responses in order to decide the right machining parameters. The most common wear pattern used in AMMNC machining is flank wear. Ra decreases with increasing cutting speed, but increases with increasing feed rate and depth of cut. This is because the machining mechanism is more stable and has less chatter at high speeds relative to low speeds. The findings lead to an understanding of the impact of the combination micro- and nanoparticles on the properties of AMMNC and its machinability.

Kumar et al. (2017) analyzed and established the effect of cutting conditions on cutting forces and surface roughness when turning nanocomposites (Al2219/nB_4C/MoS2) with cutting carbide- coated and TiN-coated carbide inserts. It is also critical to understand the relationship between cutting forces, workpiece properties, and cutting effects.

Suresh et al. (2014) optimized the machining parameters of an AlSiCGr hybrid composite turning using a gray fuzzy algorithm. Process parameters such as cutting speed, feed rate, depth of cut, and reinforcement weight fraction all have an impact on surface roughness, material removal rate, and flank wear. An optimal set of input turning parameters are required to reduce tool surface roughness and flank wear while increasing material removal rate. The gray fuzzy logic approach provides a viable solution for determining the optimal setting of multi-performance machining parameters.

Palanikumar et al. (2006) illustrated the machining of GFRP composites with various fiber orientations. Machining experiments were carried out on a lathe with a coated carbide cutting tool. Experiment designs were carried out using Taguchi's orthogonal array with the cutting conditions prefixed. Workpiece fiber orientation, cutting pace, feed rate, depth of cut, and machining time were optimized with several reaction characteristics in mind, including material removal rate, tool wear, surface roughness, and specific cutting strain. gray relational regression was used for optimization.

Joardar et al. (2014) investigated the influence of cutting parameters on cutting forces in the straight turning of LM6 Al/SiCP (MMC) in dry cutting conditions using response surface methodology (RSM). Cutting speed, depth of cut, and SiCP wt% in the metal matrix were selected as the affecting parameters, and a face-centered composite configuration was used to gather experimental data and analyze the influence of these parameters on cutting forces. RSM was used to construct a second-order model between the independent parameters and the cutting forces. The study revealed that the expected and experimental values were very similar.

Using the Taguchi method, the impact of cutting speed, feed rate, and depth of cut on cutting force and surface roughness in the turning of AA2219-TiB_2/ ZrB2 in situ MMC was investigated by Mahamani (2014). To examine the impact of turning parameters, an L27 orthogonal sequence, response graph, and analysis of variance were used.

Kishore et al. (2014) assessed the effects of cutting power, surface roughness, and flank wear for various levels of cutting speed, feed, and depth of cut for Al6061, Al6061–2%TiC, and Al6061–4%TiC. According to machinability reports, cutting force decreased as cutting speed increased. As the feed rate and depth of cut increased, so did the importance of surface roughness decreases. The higher the cutting speed, the higher the cutting temperature, and the softer the tool cutting tip.

The goal of Rao et al. (2014) was to look into the machining parameters that produced the best surface roughness. Composites present a significant machining problem due to their brittleness and higher bulk hardness compared to parent alloys. Machining MMCs at normal parameters results in cutting tool breakage due to BUE structure and the presence of fly ash particles. Surface texture degradation is caused by the formation of a BUE or a built-up layer.

As per the research, most studies focused on either tool wear or surface roughness. A comprehensive analysis of the performance of nanocomposite processing is very rare. Many studies on tool wear and surface integrity have revealed that abrasion by the rough reinforcing phase of the nanocomposite is the primary cause of tool wear. As a result, comprehensive research into nanocomposite turning is needed. Experiments were carried out using various design of experiment methods such as the Taguchi method, etc., and models for predicting success in the machining of nanocomposites were developed using RSM. Finally, optimization is accomplished to reduce the expense and time required to machine nanocomposites.

2.2.2 Milling

Milling is a processing operation where a rotary movement is performed by the cutting tool and a straight movement is performed by the workpiece. The system

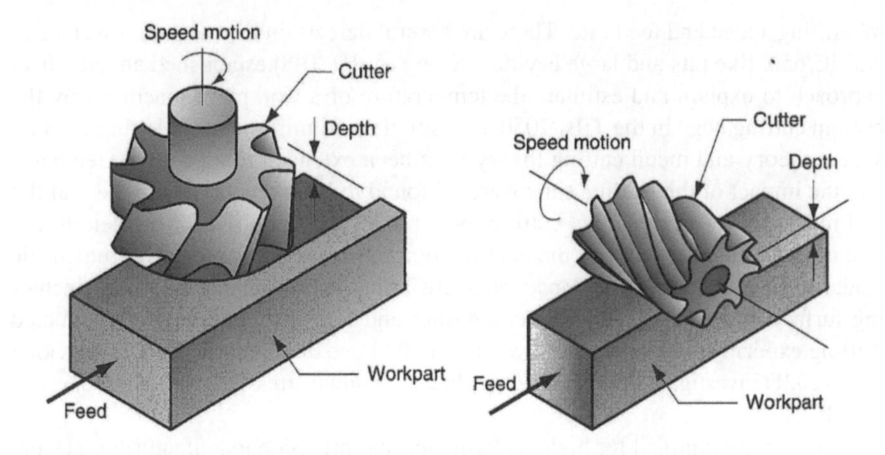

FIGURE 2.2 Schematic illustration of two basic milling operations.

is mostly utilized by multi-toothed milling cutters or end mills to operate exterior plates, slots, and contoured surfaces. For cutting surfaces of revolution, cutting metals, threading and cutting gears, milling cutters are also necessary. The method of engraving out every tooth is periodically interrupted during milling, and the cross section of the undeformed chip is not constant. The direction of cutter rotation distinguishes two types of milling: up milling and down milling, depicted in Figure 2.2.

The need for nanocomposite machining is growing and so is the demand for machining. In particular, milling is commonly used to meet the need for larger and comparatively thicker composite parts requiring extensive side milling and face milling, especially in jet engine casings. Hole making usually involves milling in order to meet the stringent stability requirements by cutting the hollow under the form until the final diameter is matched.

Xiong et al. (2016) studied the influences of milling parameters, tool wear, and workpiece material characteristics on the surface integrity during milling of in situ TiB_2-particle-reinforced Al matrix composites.

The machinability of magnesium metal matrix composites (Mg-MMCs) with large SiC fractions were investigated experimentally by Li et al. (2010) for this study. The cutting force was observed mostly with increase in spindle velocity, feed rate, and/or volume fraction. The authors suggest more detailed theoretical analysis is needed to better grasp the impact of volume fractions on the quality of machining and of the machined surface. Cutting forces, surface morphology, and surface roughness were measured to understand the machinability of four different materials. Milling was carried out at different feed rates and spindle speeds chosen according to the design of experiment (DOE) method (Li et al. 2013). Surface integrity of the high-speed milling of SiC/Al composites and subsequent unreinforced matrix alloy Al 6063 with PCD was established by Wang et al. (2013); they indicated that the most important parameter for surface roughness was milling speed, followed by the interaction between feed rate and milling speed, then the feed rate. The surface roughness of Al/SiC/65p increased with a rise in the feed volume. In terms of residual stress, the axial depth of the cut has the largest effect on residual surface stress, accompanied by milling speed and feed rate. There are several defects on the machined surface of Al/SiC/65p, like pits and large cavities. Xiong et al. (2018) established an analytical approach to explain and estimate the temperature of a workpiece machined by the bottom cutting edge in the TiB_2/7050Al in situ flat end milling unit. Moving thermal source theory and metal cutting theory have been extended to modeling. Geometry and the impact of the heating time were all found to be compatible with the real flat end milling facts. A dynamic, cutting force model was acquired for the determination of dynamic heat flux and the heat fraction ratio was experimentally studied. Jie et al. (2020) investigated the impact of in situ TiB_2 particles on machinability, including surface integrity of TiB_2/2024 composites and TiB_2/7075 composites, conducted milling experiments related to conventional 2024 and 7075 aluminum alloys. Xiong et al. (2021) investigated the surface roughness, residual stress, and fatigue properties of TiB_2/Al.

MMCs are examined for high performance and precision manufacturing. 2D and 3D surface roughness were evaluated, and the influence of operating conditions along the machined surface roughness was observed. The residual strain in the machined surface and subsurface too was studied in detail by assessing the influence of cutting force and temperature. Finally, the fatigue properties and fracture surface of the machined specimens were studied and a fatigue life prediction model based on 3D surface roughness was developed.

Liu et al. (2019) conducted an experimental investigation to study the impact of cutting parameters on surface roughness in the TiB_2/7050Al in situ cutting of MMCs. The cutting depth and feed rate were found to cause a significant impact on 3D roughness. With higher feed rate, cutting depth, or distance, 3D roughness considerably increased, and declined as the cutting speed increased. The significant difference between the surface roughness measured along its cutting and feed directions demonstrated the great influence of the measurement direction. Shihab et al. (2020) examined the effect of wt.% of reinforcement particles, i.e. SiC/Gr as variable and its influence on surface integrity (surface roughness and micro hardness) along with other end milling variables such as spindle speed, feed rate, and cutting width. They, too, used a hybrid desirable approach to multi-objective optimization. It has been noticed that, apart from spindle speed and feed rate, the wt.% of reinforced materials has a direct effect on surface integrity. Premnath (2015) studied a similar RSM model with optimization of the desirability function, taking into account different milling parameters such as feed, reducing speed, cutting depth, and weight (wt) fraction of alumina (Al_2O_3). Optimum machining parameters based on the experimental findings have shown that lower cutting strength, surface roughness, and tool wear can be accomplished by applying a combination of higher cutting speed, low feed, lower cutting depth, and higher wt fraction of alumina while face milling hybrid composites using polycrystalline diamond inserts.

Wagih et al. (2018) proposed a model for choosing the optimum milling parameters for achieving the fastest rate of particle size reduction. The theoretical model used in the optimization process is tested by applying its estimation to experimental findings from the literature as well as other results obtained during this work. The effects of milling speed and ball size have been studied to demonstrate the impact of these parameters on milling performance. Finally, using the proposed analytical model, the optimal milling time, ball size, and milling speed are calculated. To verify the model's validity, a Cu-5%ZrO2 nanocomposite was manufactured using the mechanical alloying technique and mechanically characterized.

Abdellahi et al. (2014) obtained a precise approximation of the milling parameters in order to maximize the energy transmitted to the synthesized nanopowders. Maximum energy represents the leading stage of high-energy ball milling to reduce nanocomposites synthesis time. The mechanical alloying method was modeled and optimized using powerful techniques such as gene expression programming and artificial bee colony algorithms in this article. Milling parameters such as reinforcement, form, and amount of process control agent, mill type, type of vial, type of ball, vial spinning rate, ball to powder weight ratio, milling environment, and milling time were designed to obtain the smallest grain size.

Farahnakian et al. (2011) studied the effects of nano clay and milling parameters (spindle speed and feed rate) on the machinability of polyamide-6 nanocomposites, as well as the cutting force and surface roughness of nanocomposite materials using a high-speed steel end mill. The cutting force and surface roughness were modeled independently using a particle swarm optimization-based neural network, and the capacity modeling was compared to that of a traditional neural network.

Sheikhzadeh and Sanjabi (2012) described the effects of milling time on the structure and properties of a stainless steel/30% titanium carbide nanocomposite produced by planetary milling in an argon atmosphere with stainless steel 316 and titanium carbide powders. The main goal of this research was to achieve a fine distribution of TiC nanoparticles in the steel matrix. Scanning electron microscopy (SEM), X-ray diffraction (XRD), transmission electron microscopy (TEM), and optical microscopy have been used to study the microstructure development of milled powders and nanocomposites. The XRD results showed that after 40 h of milling, the micron-size TiC had improved to nano size. The crystal size of TiC was determined to be about 12 nm during this milling period. At 40 h of milling, SEM images showed a well-distributed particle distribution. The titanium carbide powders with grain sizes less than 50 nm were also found inside of the stainless-steel matrix powder, as per TEM images. The powders manufactured at a milling time of 40 h recorded the highest density of composite. In its best condition, the composite had a hardness of about 67 Hardness Rockwell C (HRC).

Tabandeh-Khorshid et al. (2016) investigated the hardness of pure Al and Al reinforced with Al_2O_3 nanoparticles or graphene nanoplatelets (GNPs) formed by a room temperature ethanol milling procedure. The aims of this study were to see if room temperature milling in ethanol preceded by a low-temperature drying treatment will produce NC Al and NC Al MMNCs with grain sizes compared to cryomilled materials, and whether strengthening mechanisms are involved in these materials and whether the inverse Hall–Petch behavior observed in NC Al with grain sizes less than 110 nm is dependent on the reinforcement form. The average grain sizes of the processed Al_2O_3 samples were unaffected by the concentration of Al_2O_3 reinforcements. Grain boundary strengthening is the predominant strengthening mechanism in Al-Al_2O_3 NC MMNCs processed in this sample. The influence of reinforcement grain boundary pinning through processing factors for the bulk of the strengthening.

As a result, regulating grain size has been the most important factor in increasing the Al-Al_2O_3 samples' strength and hardness.

Zinati and Razfar (2014) investigated the surface roughness of a polyamide (PA) 6/MWCNT composite through end milling. A friction stir method was used to create a PA 6/MWCNT nanocomposite. The nanocomposite was however end-milled using a CNC milling machine with many cutting parameters. The machining performance of the PA 6/MWCNT composite was found to be very similar to that of metals, and the roughness of the PA 6/MWCNT composite matches the general law. In ANOVA, feed is found to be the most important factor assessing surface roughness. A combination of lower feed rate and higher spindle speed results in optimal surface consistency.

Santos et al. (2016), in their research, have tried to determine the shear bond strength of veneering porcelain to zirconia substrates that had been changed by a CNC-milling process or by coating zircon with a composite interlayer. As zirconia particles were applied to porcelain as a strengthening process, the mechanical properties of the porcelain improved significantly compared to feldspar-based porcelain.

In this section, the fundamental properties of traditionally milled nanocomposite materials were recalled. Studies to estimate responses like shear forces and roughness, etc., were discussed. Research on conventional methods of removing materials in nanocomposites is still in the early stages. Machining nanocomposites is a new challenge. To meet industrial demand, a variety of actions, such as materials, equipment, and working techniques, need to be put in place. The potential of nanomotors is undermined by a lack of consciousness about the machinability of nanocomposites along with a generic aircraft. The heat generated during cutting will damage or melt the matrix, affecting chip removal and tool wear. Effective heat transfer and material removal during dry cutting is an architectural problem of the machine tool.

2.2.3 Drilling

Drilling is a machining technique that uses twist drills to create a circular cross section hole. it involves two basic motions: the primary rotary motion and the auxiliary linear feed motion. Drilling is the most common material removal operation in metals and composites machining among various other machining operations. Drilling is done with conventional upright drilling machines, milling machines, and a wide range of specialized machines. The increased mechanical properties of nanocomposites have led to similar increase in use across a variety of industries. Higher strength-to-weight ratio, better corrosion resistance, fatigue resistance, and thermal expansion compared to metals have increased nanocomposites' applications. Moreover, by designing for specific use and components to nano-composite and directional properties, material properties can be tailored to meet specific requirements. Because of the anisotropic and heterogeneous nature of these products, during drilling, some machining damage occurs, which affects their performance and life span. Of these damages, delamination is the worst problem, which not only affects the fatigue life of the material, but may also undermine the surface finish. In recent years, consumer expectations increasingly shifted the emphasis to the development of better machining technology, posing significant challenges to manufacturers. Because of the prevalence of riveting and fastening in mechanical components and systems, hole creation is one of the most important nanocomposite supplementary metalwork techniques. Many nontraditional methods, such as laser beam drilling, water-jet drilling (with or without abrasive additive), ultrasonic drilling, and electro-discharge machining, were known as alternative approaches. Modern drilling, on the other hand, is still widely used for economic reasons. Since nanocomposites are a pretty recent material, a couple of experiments on nanocomposites drilling have indeed been conducted. Nevertheless, mechanical drilling has been shown to unfavorably damage composite materials, so it is important to assess the impact of drilling on nanocomposites (Starost and Njuguna 2014). It has been identified that numerous parameters during the drilling operation can influence the drilling factors and lead to material damage. The influence of machining parameters, tool geometry, tool materials, and tool types on cutting force generation is very important in a drilling operation. Drilling-induced damage, including delamination, burrs, swelling, splintering, and fiber pull-out, is obvious while machining nanocomposites. Delamination damage is a significant concern because it poses a direct danger to structural reliability. Numerous techniques are employed to investigate the drilled consistency over tool geometry, materials, and drilling parameters.

Kumar and Singh (2019) investigated the delamination of machined holes and their surface consistency in the drilling of epoxy/carbon-enhanced polymer nanocomposites with MWCNTs. Delamination has basically two mechanisms: push-out and peel-up. Push-out delamination is more common in drilling of CFRP composite laminates than peel-up delamination. Investigations were performed with an 8-mm-diameter AlTiN-coated solid carbide drill to investigate the effect of MWCNT wt%, spindle speed, and feed rate as process parameters on thrust force, torque, delamination (at the entrance and exit points), and surface roughness. For the preparation of MWCNT-doped CFRP nanocomposites, three different weight percentages of MWCNTs were mixed in the polymer matrix, namely 0.5%, 1.0%, and 1.5%. MWCNT had inner and outer diameters of 1–5 nm and 5–20 nm, respectively, and a length of 10 mm with a purity of around 98%. Carbon-fiber-woven roving of 600 gsm was used as reinforcement. Matrix materials included Lapox L-12 bisphenol-based epoxy resin and Lapox K-6 AH-312 epoxy hardener. Flexural strength and interlaminar shear strength (ILSS) tests were carried out to investigate the mechanical behavior of the fabricated laminates. Drilling of the fabricated specimens was done with EMCO Concept Mill 250 CNC machine at a maximum spindle speed of 10,000 r/min and 7 kW of spindle power. To measure the thrust force and torque signal, a rotating 4-component dynamometer (Kistler, type-9123C) was used. The drill tool was attached to the dynamometer, while the work piece specimens (100 mm × 100 mm × 4 mm) were held by a fixture for protecting it from vibration during the process. As the content of MWCNT in polymer matrix is increased, flexural strength and ILSS increase. It has been seen that with an increase in MWCNT wt% both thrust force and torque decreases. Delamination increases with feed rate as thrust force increases accordingly. However, with an increase in spindle speed and MWCNT wt%, delamination at the entrance and the exit sides decreases. Addition of MWCNT to the polymer matrix leads to a reduction in surface roughness.

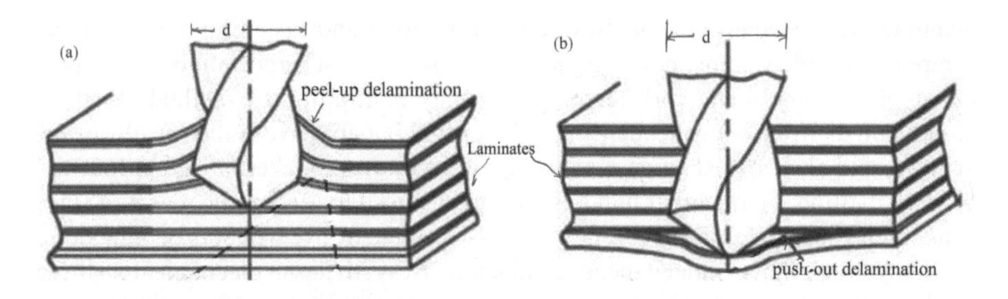

FIGURE 2.3 Schematic illustration of delamination mechanism: (a) peel-up delamination and (b) push-out delamination. (From Geng et al. (2019).)

Drilling is the most widely used machining operation for composite laminates due to the need to join structures. Several parameters have been established during the drilling process that can affect the drilling factors and material damage. In drilling operations, the effect of machining parameters, tool geometry, tool materials, and tool styles on cutting force generation is crucial.

Drilling was performed on all samples, both with and without MWCNTs, to determine the latter's effect on drilling-induced damage and surface finish (Figure 2.3).

Chakaravarthy et al. (2020), in his research, found that a cryogenic machining of nanoscale-SiC-reinforced aluminum matrix composites gives experimental results as well as a correlation of its performance with dry machining and minimum quantity lubrication. The drilling experiments were conducted on a vertical machining center operated by a CNC, with carbide drills of 10 mm diameter and cutting point angles of 90°, 118°, and 135°. Experiments have been designed using the reaction surface technique and the Box–Behnken system. Teaching-Learning-Based Optimization is used to refine drilling parameters such as spindle speed, feed rate, wt% of nano SiC, and cutting angle.

The objective of Heidary et al. (2018) was to investigate the effect of nanomaterials and drilling parameters (feed rate, cutting speed, and drill diameter) on drilling thrust force, delamination size, and residual flexural strength of E-glass epoxy/MWCNT laminates. The Taguchi method with Gray relational analysis has been used to optimize the drilling process with multiple performance characteristics. As per the study results, feed rate has the greatest impact on the thrust force and delamination factor, followed by spindle speed. Nevertheless, nano content has the greatest influence on residual flexural strength, followed by feed rate.

D'Orazio et al. (2017) used two CFRP laminates at the top and bottom of the specimen, with an AA7075 core plate sandwiched between them. The stack measured 300 mm in length, 210 mm in width, and was 26.1 mm thick in aluminum and 2.8 mm thick in CFRP laminates. The 5-axis CNC machining center JOMACH 141 was used for the drilling process. The diameter of the hole was chosen to meet the specifications of the automotive industry, which was 6.8 mm. Diamond-like carbon (DLC) and TiAlN were the two types of coatings used on the tungsten carbide drill equipment. With each coated tool, 170 holes were drilled in the CFRP/AA7075 stacks. To determine tool wear on the flank surface, the image analysis device LASEZ (Leica Application Suite) was used. The dynamometer setup was used to measure the vertical force (F) produced during drilling. LASEZ was used to find the delamination factor and calculate the delamination region. The hole diameters were measured three times along the length of the stack with the Smart Scope ZIP300 coordinate measuring unit. The DLC-coated drill has been found to have substantially less wear than the TiAlN-coated drill. On both AA7075 and CFRP, the peak thrust force increases as the number of holes increases. The number of holes raises the delamination factor, and it was discovered that much less delamination occurs when using a DLC-coated tool. Owing to tool wear, hole diameter decreases as the number of holes per layer increases, and the exit hole diameter is much smaller than the entry hole. A third-degree polynomial regression model was used.

Vital process parameters such as MWCNT wt%, spindle speed, feed rate, and tool materials have indeed been optimized by Kharwar and Verma (2020), who worked to reduce surface roughness, thrust power, and torque force. Surface roughness is significant during the drilling of polymer composite material when the component is subjected to fatigue load, rivet fastening, precise fits and holes, and so on. The thrust force and torque force produced during drilling have a significant impact on the surface quality of the hole and the damage around the surface. Cutting forces are critical parameters that must be kept to a minimum when drilling. To evaluate the application potential, a comparative analysis of GRA and GRA-PCA hybrid modules was conducted, which revealed that hybrid GRA-PCA is more capable than conventional GRA in terms of less average error (5.037%) and total gain of 13.99% in evaluation values.

Celik et al. (2019) has converted graphene (G) to graphene oxide (GO) nanoparticles, which were then mixed with epoxy in various ratios to establish the tensile strength of nanocomposites. By hand lay-up, the reinforcing ratio of nanocomposites with the maximum tensile strength, epoxy with G and GO, and nonreinforced epoxy is taken into account. Consequently, the CFRP composite as well as the G/CFRP and GO/CFRP nanocomposites were developed. The effects of G and GO on the cutting force, cutting torque, temperature, and delamination factor of the fabricated nanocomposites were studied. It was discovered that GO was successfully synthesized, and that G and GO had a beneficial effect on tensile strength. GO demonstrated a more powerful function on tensile power than G.

Juliyana and Prakash (2020) used stir casting to create a novel composite made of LM5 aluminum alloy bonded with zirconia in three different wt% (3%, 6%, and 9%). The Taguchi method was used to conduct the experiments with the best drilling parameters. By drilling in a CNC machine, the effect of input parameters such as feed rate, spindle speed, drill material, and percentage of reinforcement on thrust force is determined. The most important input parameters for achieving minimum thrust force are spindle speed and feed rate.

Singh and Kumar (2018) used MWCNTs to improve the mechanical properties of laminates. As the feed rate and drilled diameter increase, the surface roughness of the holes increases. Surface roughness is influenced more by MWCNTs. As a result, smaller holes can be drilled instead of larger ones. The delamination factor decreases as the spindle speed increases, but increases as the drill diameter increases. The strength and fracture toughness of GFRP nanocomposites are improved when MWCNTs are added to the matrix, as a result of less delamination on the entry side and more on the exit side. The findings show that increasing the MWCNT wt% decreases the surface roughness of the holes that are drilled, but the delamination factors decrease as the drilling speed increases.

Kharwar et al. (2020b) highlight the optimization of drilling parameters using a hybrid method of multi-objective optimization based on ratio analysis (MOORA) and PCA. The aim of drilling is to minimize surface roughness, torque, thrust power, and delamination at the entry and exit points. A mixture of high speed and low feed rate is critical when machining polymer nanocomposites. According to the report, the optimal MOORA–PCA hybrid solution environment is 1.0 wt%, 1000 rpm, 50 mm/min, and TiAlN tool materials, which results in a significant improvement in the overall MOORA-PCA measured value. The samples used in the analysis were drilled on a CNC machine model no. BMV 35 TC20 with three distinct wt% of MWCNT.

Kumar et al. (2020) proposed a new method for reducing delamination and thrust force developed during the drilling of polymer nanocomposites reinforced with graphene oxide/carbon fiber. The specimen was created using graphene oxide (Mesh size 200), CFRP prepregs unidirectional (UD), and epoxy resin. Drilling was done with TiAlN-coated SiC drill equipment with a 5 mm diameter. The increased feed rate raises delamination value, which is mostly due to the chisel tip of the drill bit. The spindle speed has a beneficial effect on delamination and can soften the composite matrix. Since high-speed machining raises the temperature between the machining interface (tool cutting edge and workpiece), less cutting force is needed. The cumulative effect of feed rate and speed greatly increases delamination in polymer composites.

The nanocomposites were created using a bisphenol-A-based epoxy by Kharwar et al. (2020a). They were then solution-cast with MWCNTs with average diameters of 10–15 nm. Drilling parameters including spindle speed, feed rate, and tool material were considered. When drilling the nanomaterial, the optimum mix increases surface roughness by 18.24% and thrust force by 26.13%. Because of the improved thermal conductivity of polymer composites, the thermal softening of the matrix phase is minimized. It prohibits surface damage and faults from occurring during machining. The PCA-embedded CoCoSo solution is a single exponentially weighted product model with a modest additive weighting. It may be due to a combination of consensus solutions for the best drilling method for a given material. The neural network is made up of input layers, a hidden layer, and an output layer, all of which are interconnected by neurons with their own weights and biases. As a consequence, even at high cutting speeds, it is easy to drill a nanocomposite.

Raja et al. (2020) created AA7075/BN-reinforced MMCs using a conventional liquid metallurgy technique. As a reinforcing material, various wt% of BN ceramic powder were used. Drilling events were carried out in a dry atmosphere using a CNC machining center. Material removal rate has been improved by optimizing effective machining parameters to achieve higher thrust force and surface roughness. The experiment was carried out using a 10 mm solid carbide drill bit. Raising the feed rate and wt% of the reinforcement greatly increased thrust force and surface roughness. Low speed (1500 rpm) and high feed (0.3 mm/min) combinations provided more thrust force than high speed and low feed combinations.

Ragunath et al. (2017) investigated drilling with nano-clay particle material, natural fiber, and GRFP composites. The study's key aim was to estimate optimum levels of process parameters and minimize delamination. Increased filler content in fiber materials improves the delamination aspect. Excess filler material causes composites to lose strength, and the study discovered that this increased the amount of filler material required to improve drilling performance.

Nanoparticle reinforcements are critical for improving the machinability of nanocomposite materials (Burak Kaybal et al. 2020). The influence of BN reinforcement on the machinability of composite materials is a relatively recent research subject. The role of BN in the drilling efficiency of carbon fiber epoxy nanocomposites is demonstrated. In addition, the effects of cutting speed, feed rate, and surface roughness on drilling efficiency were studied. The study was conducted on a Mazak Variaxis 500 CNC machining center. The thrust force is increased by BN nanoparticles, which have a tight interface and interlaminar bonding due to their ceramic composition and hexagonal crystal structure. The cutting speed and feed rate were found to be the most important drilling parameters. At exit surfaces, feed appears to decrease as cutting speed increases. The researchers suggest it may also be important to use a particular drill bit geometry to help minimize delamination when drilling nanoparticles with hybrid composite material included in the composite material. Nanoparticles in the nanocomposite also improved the shear and bending resistance between laminates.

Drilling parameters for surface roughness and circularity error when machining MWCNT-reinforced epoxy nanocomposites are being investigated by Verma et al. (2021). The solution casting method was used to create MWCNT-reinforced epoxy nanocomposites. To strengthen the epoxy, three different wt% of MWCNT (0.5,1.0,1.5) were used in this procedure (Lapox, L12). The nano reinforcement measures 15 mm in length and has a diameter of 10–15 nm. They made a nanocomposites sample that measured $10 \times 10 \times 7$ mm in diameter. The drilling experiment was carried out with drill bits made of three different materials (high-speed steel, carbide, and TiAlN) with a diameter of 5 mm and a 118° point angle and a 30° helix angle. Drilling parameters were measured at three levels: MWCNT wt% (0.5, 1.0, 1.5), spindle speed (500, 1000, 1500 rpm), feed rate (50, 100, 150 mm/min), and tool content (HSS, carbide, TiAlN). The effect of process parameters on surface roughness and circularity error was investigated using ANOVA. Surface roughness rises with higher feed rates and falls with higher spindle speeds. Up to a certain amount of reinforcement, the effect of MWCNT wt% increases surface roughness. At 1.0 MWCNT wt%, surface roughness was found to be lower. Due to the dense network of MWCNT and epoxy, the higher the MWCNT wt%, the rougher the floor. A combination of higher spindle speed and lower feed rate is optimal for lower surface roughness values. Since the circularity error is primarily controlled by the cutting mechanism, the MWCNT wt% has a major effect on cutting behavior, decreasing with spindle speed and increasing with feed rate. The cutting mechanism is influenced by the tool material, which has a different effect during machining due to its different hardness. Tool content is another criterion for controlling machining activity, according to experimental findings.

Reddy et al. (2020) used gray relational analysis to optimize the drilling constraints simultaneously during the drilling of an Al6063/TiC composite. Drilling experiments were performed on Al-6063 drill bits with a diameter of 50 μm and a titanium carbide (TiC) content of 15%. The composite used in the study was Al6063 with 15% TiC MMC, had a thickness of 10 mm, and was processed using the liquid processing method. Three limiting factors were considered to evaluate surface roughness, cutting strength, and drilling temperature: helix angle, spindle speed, and feed rate, as well as HSS drill bits with a diameter of 10 mm. A pyrometer was also used to measure the temperature of the drilling. The temperature of the drilling tool–workpiece interface was measured with a pyrometer. The experimental findings were interpreted using GRA and ANOVA for the simultaneous optimization and effect of drilling parameters on responses. The optimum combination of drilling parameters discovered by GRA in this analysis is a higher helix angle (40°), a medium spindle speed (500 rpm), and a lower feed (0.1 mm/rev). According to the GRG's ANOVA, the most influencing parameter is helix angle, which contributes 63.2%, followed by feed, which contributes 34%.

Xu et al. (2019) conducted an experimental study of the machinability of one kind of high-strength T800/X850 CFRP that is typical in aircraft components. The thrust force for both the brad spur and twist drills has evidently improved due to the increased spindle rpm. The brad spur drill provides the lowest thrust forces under the measured drilling conditions due to its high point angle and superior diamond coating, which reduces friction at the tool–workpiece interface. It is followed by the twist drill and dagger drill. This implies that the drilling parameters have a negative impact on the composite surface finish, and that a combination of low spindle speeds and low feed rates will allow for damage-free drilling of high-strength CFRP laminates.

Significant efforts have been made to improve knowledge of the phenomenon related to the mechanism of drilling-induced delamination. Distinctive drill bits, support plates, pre-drilled pilot holes, vibration-assisted twist drilling, and high-speed drilling are also used to increase the accuracy of drilled holes. There are analytical models of delamination and thrust force for drilling composite laminates, but they are insufficient to highlight the physical significance of drilling composite laminates. During composite laminate drilling, the feed rate seems to have the greatest influence on delamination, thrust power, and tool wear.

2.2.4 Grinding

Grinding is primarily a technique for removing materials under which a comparatively large number of widely scattered, excessively hard and stable abrasive, very complex, shape and geometry, instead of a single or a few evenly spaced and oriented cutting edges of equal and well-defined geometry, as in any traditional machining, achieve material removal in the form of small chips. Figure 2.4 schematically shows the grinding tests setup.

Nanoparticle composites are usually reinforced by types of particles reinforced very high hardness alloys, such as titanium alloy, B_4C, and the reinforced phase of the particle has been used to improve the mechanical properties due to the inclusion

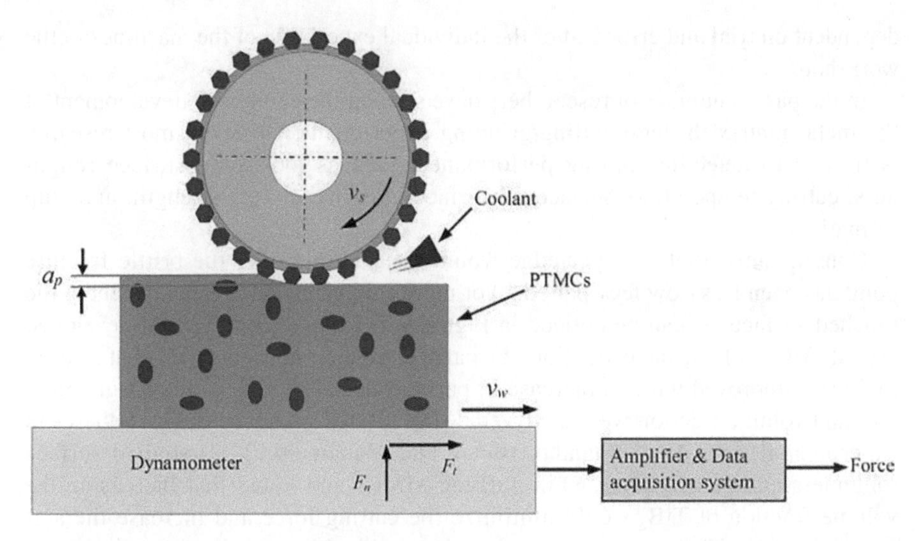

FIGURE 2.4 Schematic illustration of grinding experimental setup (Adapted from Liu et al. (2018).)

of the reinforced phase of the particles, which have normally had superior properties such as high specific stiffness and high specificity. As a result, there is a high prospect for their deployment in specialized industries such as the aviation industry. Over the last few years, grinding with abrasive tools to achieve high performance and precision machining of these nanocomposites has been studied for this purpose. Based on research, it is also known that the hard particles present in the nanocomposite can deteriorate because of the extreme wear of the tool and the integrity of the surface, thereby inhibiting the performance and quality of the surface.

Grinding is the most widely used process for machining hard and brittle nanocomposites. The behavior of nanomaterials and their relationships are currently being studied intensively. Grinding, indentation, and, in general, all processes involving the use of abrasive grit to interact with the surface of the material are still uncertain.

Analogous to machinability, the term grindability is used, which corresponds to easy grinding and is primarily assessed by the magnitude of grinding forces and the specific energy requirements. Grind temperature affects the quality and life of the product, surface completeness, including surface finish, residual stress, and microcracks in surface and surface areas, life span, and grinding ratios.

The appropriate selection of the grinding wheel and process parameters and the application of appropriate grinding fluids will improve the grinding capacity of any material. A rigid and durable design with great stiffness in the grinding machines is essential. In recent decades, there have been a series of revolutionary changes and innovations to meet the growing demand for higher output and crushing quality.

The process of grinding nanocomposites is an important subject for research, and its physics is well accepted. A description of the problems experienced during this process was also necessary, but it was not sufficient to help optimize the process due to the lack of realistic quantitative relationships or analytical models that might be mathematically optimized. Consequently, the selection of process variables was dependent on trial and error and/or the individual experience of the machines in the workshop.

In the past, a number of researchers have studied the composite development of the metal matrix through cutting/grinding experiments. However, most research is limited to detecting cutting performance, such as tool wear, surface roughness, cutting temperature, surface/subsurface integrity, cutting strength, and chip formation.

Consequently, tool feed per edge would conclude whether the brittle fracture point has been less (low feed per edge) or more (high feed per edge) adjacent to the finished surface, as can be noticed in Figure 2.5. Kannan et al. (2006) developed 6061 MMC with variable fractions of particle volume and observed that cutting tool wear improved with an increase in particle size. The effects of strengthening size and volume fraction were analyzed. After all, the proposed design adheres to experimental wear data. Anandakrishnan and Mahamani (2011) studied surface roughness and cutting force while turning MMCs and found that increasing the volume fraction of TiB_2 would minimize the cutting force and increase the surface roughness. The results of variations in machinability, such as cutting speed, feed speed, cutting depth, sidewall wear, cutting resistance, and surface roughness, were analyzed during turning operations. The high TiB_2 hardening rate increases tool wear and decreases cutting strength. Increased feed speed increases sidewall wear and shear strength. A finite element model for high-speed grinding of particulate reinforced titanium matrix composites has been developed by Liu et al. (2018). The grinding force has different properties where the alloy matrix and the reinforcement particle have been extracted, respectively. The undeformed thickness of the chip has a greater effect on the creation of machined surface defects. The findings of the simulation were well correlated with the experiment, according to the authors' experiment.

Zhong and Hung (2002) investigated the results of experiments on surface/subsurface integrity of MMCs ground with a SiC wheel and a diamond wheel. In his work, he describes the results of research obtained by grinding aluminum-based MMC reinforced with Al_2O_3 particles using grinding wheels having SiC in the die and diamonds in a resin-bound die. The concerns addressed include surface roughness, low feed per edge ductile mode milling process, high feed per edge brittle mode milling process, grinding force, shape and scale of abrasives, grinding conditions, and consequential subsurface integrity.

Kwak and Kim (2008) examined the fabrication and grinding of aluminum-based MMCs reinforced with SiC particles. The mechanical properties of the MMCs developed and the grinding effects of the tests were examined and the effect of grinding parameters on the surface roughness and grinding forces of the MMCs was assessed. In addition to experimental research, various numerical simulations and research on MMC processing have also been carried out. Zhu and Kishawy (2005) built up a finite element model using commercial software by combining a thermo-elasto-plastic substance model and a thermoplastic cutting tool model, with specific attention to the tension of the particles and the aluminum matrix. They did not just predict the cutting strength, but defined the distribution of stress, shear strain, and cutting temperature of the aluminum matrix and reinforced particles during the cutting process.

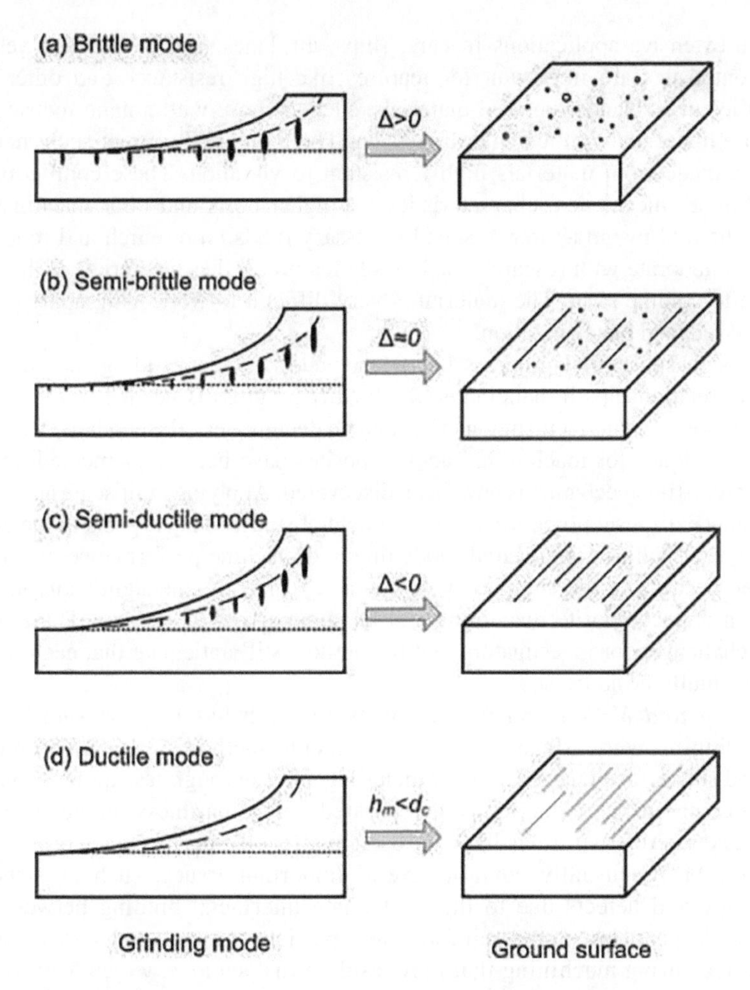

FIGURE 2.5 Four different grinding modes for horizontal surface grinding (Adapted from Gu et al. (2011).)

In brief, several studies have investigated the machining method of MMC through experimentation and simulation. Conversely, much of the work focused on the aluminum matrix cutting process composites.

2.3 CHALLENGES AND FUTURE PROSPECTS OF CONVENTIONAL MACHINING

There is no doubt that nano-sized materials have applications in many scientific and technological fields. Nanocomposites have been among the materials and alloys that are currently experiencing tremendous growth, particularly in the application process. An emerging trend toward high-quality materials, specifically in various advanced industries, will lead to an exponential increase in the use of nanocomposites in different industrial sectors. Nanocomposite materials are expected to have extensive applications in cars, ships, airplanes, and even space vehicles. Nanocomposites are important for features like high resistance and other fairly decent features. Nanostructured materials, such as those with a nano-metric thickness, are materials with a single dimension. The tools used to treat such materials must be made from materials highly resistant to vibration. These conflicting features (higher mechanical characteristics vs. higher costs and poor machinability) enable the tool manufacturer to spend necessary funds on research and innovation and to collaborate with researchers. The key feature of these materials is their high surface-to-volume ratio. The material is very difficult to work with, apart from its cost and reduced mechanization.

Although substantial progress has been made in understanding the structure–property relationships in nanomaterials, further progress is needed in the areas of nanocomposite manufacturing using suitable techniques. Researchers show that many techniques for machining nanocomposites have been implemented and that many interesting mechanisms have been discovered. Applying multiscale approaches to substance disposal processes is a recent field of study. Many aspects of the current approaches, such as complicated modelling and machine performance, need to be changed because of the complexity of systems. Currently, the significant advancements in nanocomposite machining can be summarized as follows. These are all tough challenges for investigators, but the authors still anticipate that each of them will eventually be addressed.

Turning and Milling: Surface integrity research has become significant in the machining world since the most influencing method of manufactured surface and subsurface characteristics, including surface roughness and morphology of producing surfaces (i.e., defects generated, micro-hardness, induced residual stress, and microstructure) is likewise used to affect the performance of machined surfaces. MMCs usually contain several important issues, such as structural anisotropy and defects due to the inadequate interfacial binding between hard and ductile particles. Some reinforcement particles are so tender to press into the matrix during machining that they result in dislocations, which lead to stress hardness of the machine's surface and subsurface, and machining defects such as pits, voids, microcracks, grooves, protuberances, and matrix tearing on the machined surface.

Drilling: Research has concentrated only on the investigation of nanoparticle release in the limited studies on nanocomposite drilling. This shows that no studies are currently evaluating the material damage caused by nanocomposite drilling. The real challenge in nanocomposite drilling is the production of damage-free holes. In drilling, damage via delamination takes place on the entry/exit surfaces of the specimen. Torque as well as thrust force are often investigated by researchers to judge their correlations. Furthermore, the geometry of the drill bit, the material of the drill bit, the cutting conditions, the characteristics of the work material, and the coating are all being studied to better understand their effects on delamination. The overall goal of these studies is to remove delamination to minimize the risk of composite material components' failure during service life. The feed rate and the cutting speed are raise in the drilling parameter to have a significant impact on thrust strength and hence delamination in the drilling process. Experimenters have reported that higher feed rates trigger greater damage to the drilled hole. Some studies have hypothesized that cutting speed is indeed a significant factor in delamination formation, while other researchers have indicated minimal effects on delamination. Drilling nanocomposite laminates places particular demands on the geometry and wear resistance of the drill bit. Many such special drill bits have been developed to mitigate drilling-induced delamination. Carbide tools, coated carbide tools, and PCD tools all perform well in terms of tool wear and life.

Although no sufficient knowledge exists of many processes, the machined surface integrity of composites is still influenced by various parameters such as nanocomposite composition, tooltip radius, tool geometry and flank wear, reduced feed rate, high nose range, and cutting fluids. Which could improve the machined surface of nanocomposites.

Future challenges:

There are a few challenges that need to be overcome for the strategy on industrial applications of nanocomposites to be successfully implemented. Some challenges include:

Synthesizing nanopowders: The majority of nanocomposite fabrication techniques make use of nanopowders as feedstock materials. There is a need to improve methods that are both cost-effective and capable of processing large quantities of nanoparticles.

Handling of nanopowders: As these ultrafine powders have such a large surface area, they are prone to contamination. To prevent degradation, nanopowders should be treated and managed in a novel way. Another difficult feature of nanocomposite processing is developing methods for properly treating and storing nanoparticles.

Fabrication/consolidation techniques: The promise of nanotechnology is focused on the potential to build nanostructured materials with novel properties at the "macro scale." Traditional consolidation methods include significant limitations in terms of preserving the nanostructure in the final product.

Interdisciplinary effort: The wide use of nanotechnology for public gain is completely dependent on the active involvement of all science and technology fields at all levels. Academics and administrators must create and incorporate new paradigms for teaching students and researchers.

On the basis of the abovementioned studies, it has been ascertained that their applications and machinability investigations are still constrained, primarily to the polymer matrix and metal matrix, and nanocomposites reinforced with ceramic nanoparticles, CNTs, and graphene. Complex machining methods results in interaction between variable inputs such as feed rate and cutting speed and corresponding outputs with various descriptions. The majority of research has focused on the effects of filler material, feed rate, cutting speed, chip formation, cutting force, and surface roughness. The impact of cutting tools and tool wear on the machinability of nanocomposites, on the other hand, has received very little attention. Furthermore, there has been little research into the size effect when machining nanocomposites. A general perspective on the machinability of nanocomposites has been given in this chapter. It will serve as the foundation for future studies in this area. However, the knowledge gap has facilitated increased efforts to gather quantitative data and adequate information on this area of study.

The domain of nanocomposites is relatively new, with a scientific background that includes chemistry, physics, biology, materials science, and engineering. It has the power to transform numerous industries, including space and aviation, technology, medical services, and sustainable development. The field of nanocomposites is becoming more interesting than ever because of its tremendous potential to enhance our quality of life in the future.

Nanocomposites are appropriate for the next phase in our development to satisfy emerging demands. In monolithic and microcomposite equivalents, they have better performance. They are the right candidates for overcoming the shortcomings of many existing equipment. Even at low loading reinforcements, gas barriers, and flame-related proprietors, properties like high mechanical endurance create future uses and markets for nanocomposites, leading to the emergence of the modern material age.

Future research trends may indicate that machining of ceramic matrix nanocomposites should be addressed. Furthermore, because the machining of nanocomposites has enormous industrial potential, the state of the art in this area has seen very few applications, with the majority of them still in the prototyping stage. As a result, future research may concentrate on industrial applications of nanocomposites machining.

BIBLIOGRAPHY

Abdellahi, M., Bahmanpour, H., and Bahmanpour, M. (2014). The best conditions for minimizing the synthesis time of nanocomposites during high energy ball milling: modeling and optimizing. *Ceramics International*, 40(7):9675–9692.

Agarwal, P., Kishore, A., Kumar, V., Soni, S. K., and Thomas, B. (2019). Fabrication and machinability analysis of squeeze cast Al 7075/h-BN/graphene hybrid nanocomposite. *Engineering Research Express*, 1(1):015004.

Ajithkumar, J. and Xavior, M. A. (2019). Flank and crater wear analysis during turning of Al 7075 based hybrid composites. *Materials Research Express*, 6(8):086560.

Al-Khaldi, T., Mansour, E., Gaafer, A., and Habib, S. (2019). Machinability of EPOXY/MWCNTs nanocomposites during turning operation. *Journal Homepage: www. feng. bu. edu. eg*, 1(41):26–30.

Anandakrishnan, V. and Mahamani, A. (2011). Investigations of flank wear, cutting force, and surface roughness in the machining of Al-6061–TiB2 in situ metal matrix composites produced by flux-assisted synthesis. *The International Journal of Advanced Manufacturing Technology*, 55(1–4):65–73.

Bhushan, R. K., Kumar, S., and Das, S. (2010). Effect of machining parameters on surface roughness and tool wear for 7075 al alloy sic composite. *The International Journal of Advanced Manufacturing Technology*, 50(5–8):459–469.

Burak Kaybal, H., Unuvar, A., Kaynak, Y., and Avc, A. (2020). Evaluation of boron nitride nanoparticles on delamination in drilling carbon fiber epoxy nanocomposite materials. *Journal of Composite Materials*, 54(2):215–227.

Celik, Y. H., Kilickap, E., and Kocyiffgit, N. (2019). Evaluation of drilling performances of nanocomposites reinforced with graphene and graphene oxide. *The International Journal of Advanced Manufacturing Technology*, 100(9):2371–2385.

Dandekar, C. R. and Shin, Y. C. (2012). Modeling of machining of composite materials: a review. *International Journal of Machine Tools and Manufacture*, 57:102–121.

Davim, J. P. (2001). Turning particulate metal matrix composites: experimental study of the evolution of the cutting forces, tool wear and workpiece surface roughness with the cutting time. *Proceedings of the Institution of Mechanical Engineers, Part B: Journal of Engineering Manufacture*, 215(3):371–376.

D'Orazio, A., El Mehtedi, M., Forcellese, A., Nardinocchi, A., and Simoncini, M. (2017). Tool wear and hole quality in drilling of CFRP/AA7075 stacks with DLC and nanocomposite TiAlN coated tools. *Journal of Manufacturing Processes*, 30:582–592.

El-Kady, E., Gaafer, A., Ghaith, M., Khalil, T., and Mostafa, A. (2015). The effect of machining parameters on the cutting forces, tool wear, and machined surface roughness of metal matrix nano composite material. *Advances in Materials*, 4(3):43–50.

Elango, M. and Annamalai, K. (2020). Machining parameter optimization of Al/SiC/Gr hybrid metal matrix composites using anova and grey relational analysis. *FME Transactions*, 48(1):173–179.

Farahnakian, M., Razfar, M. R., Moghri, M., and Asadnia, M. (2011). The selection of milling parameters by the PSO-based neural network modeling method. *The International Journal of Advanced Manufacturing Technology*, 57(1–4):49–60.

Geng, D., Liu, Y., Shao, Z., Lu, Z., Cai, J., Li, X., Jiang, X., and Zhang, D. (2019).Delamination formation, evaluation and suppression during drilling of composite laminates: A review. *Composite Structures*, 216:168–186.

Gu, W., Yao, Z., and Li, H. (2011). Investigation of grinding modes in horizontal surface grinding of optical glass BK7. *Journal of Materials Processing Technology*, 211(10):1629–1636.

Gupta, M. and Kumar, S. (2015). Investigation of surface roughness and MRR for turning of UD-GFRP using PCA and Taguchi method. *Engineering Science and Technology, an International Journal*, 18(1):70–81.

Heidary, H., Karimi, N. Z., and Minak, G. (2018). Investigation on delamination and flexural properties in drilling of carbon nanotube/polymer composites. *Composite Structures*, 201:112–120.

Jahanmir, S., Ramulu, M., and Koshy, P. (1999). *Machining of ceramics and composites*. Marcel Dekker.

Jie, C., Weiwei, Y., Zhenyu, Z., Yugang, L., Dong, C., Qinglong, A., Jiwei, G., Ming, C., and Haowei, W. (2020). Effects of in-situ tib2 particles on machinability and surface integrity in milling of tib2/2024 and tib2/7075 al composites. *Chinese Journal of Aeronautics*, 34(6): 110–124.

Joardar, H., Das, N., Sutradhar, G., and Singh, S. (2014). Application of response surface methodology for determining cutting force model in turning of LM6/SiCP metal matrix composite. *Measurement*, 47:452–464.

Juliyana, S. J. and Prakash, J. U. (2020). Drilling parameter optimization of metal matrix composites (LM5/ZrO2) using Taguchi technique. *Materials Today: Proceedings*, 33:3046–3050.

Kalyan Chakaravarthy, V., Rajmohan, T., Vijayan, D., Palanikumar, K., and Latha, B. (2020). Sustainable drilling performance optimization for nano SiC reinforced Al matrix composites. *Materials and Manufacturing Processes*, 35(12):1304–1312.

Kannan, S., Kishawy, H., and Balazinski, M. (2006). Flank wear progression during machining metal matrix composites. *Journal of Manufacturing Science and Engineering*, 128(3):787–791.

Kannan, C., Ramanujam, R., and Balan, A. (2018). Machinability studies on Al 7075/BN/Al2O3 squeeze cast hybrid nanocomposite under different machining environments. *Materials and Manufacturing Processes*, 33(5):587–595.

Kannan, C., Varun Chaitanya, C. H., Padala, D., Reddy, L., Ramanujam, R., and Balan, A. S. S. (2020). Machinability studies on aluminium matrix nanocomposite under the influence of mql. *Materials Today: Proceedings*, 22: 1507–1516.

Khandey, U., Ghosh, S., and Hariharan, K. (2017). Machining parameters optimization for satisfying the multiple objectives in machining of MMCs. *Materials and Manufacturing Processes*, 32(10):1082–1093.

Kharwar, P. K. and Verma, R. K. (2020). Machining performance optimization in drilling of multiwall carbon nano tube/epoxy nanocomposites using GRA-PCA hybrid approach. *Measurement*, 158:107701.

Kharwar, P. K., Verma, R. K., and Singh, A. (2020a). Neural network modeling and combined compromise solution (cocoso) method for optimization of drilling performances in polymer nanocomposites. *Journal of Thermoplastic Composite Materials*, 0892705720939165.

Kharwar, P. K., Verma, R. K., and Singh, A. (2020b). Simultaneous optimisation of quality and productivity characteristics during machining of multiwall carbon nanotube/epoxy nanocomposites. *Australian Journal of Mechanical Engineering*: 1–19. doi:10.1080/14484846.2020.1794511

Kishore, D. S. C., Rao, K. P., and Mahamani, A. (2014). Investigation of cutting force, surface roughness and flank wear in turning of in-situ Al6061-TiC metal matrix composite. *Procedia Materials Science*, 6:1040–1050.

Krishnamurthy, L., Sridhara, B., and Budan, D. A. (2007). Comparative study on the machinability aspects of aluminium silicon carbide and aluminium graphite composites. *Materials and Manufacturing Processes*, 22(7–8):903–908.

Kumar, A., Mahapatra, M., and Jha, P. (2014). Effect of machining parameters on cutting force and surface roughness of in situ Al–4.5% Cu/TiC metal matrix composites. *Measurement*, 48:325–332.

Kumar, D. and Singh, K. (2019). Investigation of delamination and surface quality of machined holes in drilling of multiwalled carbon nanotube doped epoxy/carbon fiber reinforced polymer nanocomposite. *Proceedings of the Institution of Mechanical Engineers, Part L: Journal of Materials: Design and Applications*, 233(4):647–663.

Kumar, J., Verma, R. K., and Debnath, K. (2020). A new approach to control the delamination and thrust force during drilling of polymer nanocomposites reinforced by graphene oxide/carbon fiber. *Composite Structures*, 253:112786.

Kumar, N. S., Shankar, G. S., Basavarajappa, S., and Suresh, R. (2017). Some studies on mechanical and machining characteristics of Al2219/n-B4C/MoS2 nanohybrid metal matrix composites. *Measurement*, 107:1–11.

Kumar, R. and Chauhan, S. (2015). Study on surface roughness measurement for turning of Al 7075/10/sicp and Al 7075 hybrid composites by using response surface methodology (RSM) and artificial neural networking (ann). *Measurement*, 65:166–180.

Kwak, J. and Kim, Y. (2008). Mechanical properties and grinding performance on aluminum-based metal matrix composites. *Journal of Materials Processing Technology*, 201(1–3):596–600.

Li, J., Liu, J., Liu, J., Ji, Y., and Xu, C. (2013). Experimental investigation on the machinability of SiC nano-particles reinforced magnesium nanocomposites during micro-milling processes. *International Journal of Manufacturing Research*, 8(1):64–84.

Li, J., Liu, J., and Xu, C. (2010). Machinability study of sic nano-particles reinforced magnesium nanocomposites during micro-milling processes. In *International Manufacturing Science and Engineering Conference*, 49477:391–398.

Liu, C., Ding, W., Yu, T., and Yang, C. (2018). Materials removal mechanism in highspeed grinding of particulate reinforced titanium matrix composites. *Precision Engineering*, 51:68–77.

Liu, X.-f., Wang, W.-h., Jiang, R.-s., Xiong, Y.-f., Lin, K.-y., and Zhang, Z.-f. (2019). Study on surface roughness of milling in-situ tib2 particle reinforced al matrix composites. In *ASME International Mechanical Engineering Congress and Exposition*, 59490:V012T10A020. American Society of Mechanical Engineers.

Looney, L., Monaghan, J., O'Reilly, P., and Taplin, D. (1992). The turning of an al/sic metal-matrix composite. *Journal of Materials Processing Technology*, 33(4):453–468.

Mahamani, A. (2014). Influence of process parameters on cutting force and surface roughness during turning of AA2219-TiB2/ZrB2 in-situ metal matrix composites. *Procedia Materials Science*, 6:1178–1186.

Palanikumar, K., Karunamoorthy, L., and Karthikeyan, R. (2006). Multiple performance optimization of machining parameters on the machining of GFRP composites using carbide (K10) tool. *Materials and Manufacturing Processes*, 21(8):846–852.

Poovazhgan, L. (2020). Turning experiments on Al/B4C metal matrix nanocomposites. In *Materials Science Forum*, 979:16–21. Trans Tech Publ.

Prakash, M. and Iqbal, U. M. (2018). Parametric optimization in turning of AA2014/Al2O3 nano composite for machinability assessment using sensors. In *IOP Conference Series: Materials Science and Engineering*, 402:012013. IOP Publishing.

Premnath, A. A. (2015). Studies on machining parameters while milling particle reinforced hybrid (al6061/al2o3/gr) mmc. *Particulate Science and Technology*, 33(6):682–692.

Priyadarshi, D. and Sharma, R. K. (2016a). Effect of type and percentage of reinforcement for optimization of the cutting force in turning of aluminium matrix nanocomposites using response surface methodologies. *Journal of Mechanical Science and Technology*, 30(3):1095–1101.

Priyadarshi, D. and Sharma, R. K. (2016b). Optimization for turning of al-6061-sicgr hybrid nanocomposites using response surface methodologies. *Materials and Manufacturing Processes*, 31(10):1342–1350.

Ragunath, S., Velmurugan, C., and Kannan, T. (2017). Optimization of drilling delamination behavior of GFRP/clay nano-composites using RSM and GRA methods. *Fibers and Polymers*, 18(12):2400–2409.

Raja, K., Sekar, V. C., Kumar, V. V., Ramkumar, T., and Ganeshan, P. (2020). Microstructure characterization and performance evaluation on AA7075 metal matrix composites using RSM technique. *Arabian Journal for Science and Engineering*, 45(11):9481–9495.

Rao, C. P., Bhagyashekar, M., et al. (2014). Effect of machining parameters on the surface roughness while turning particulate composites. *Procedia Engineering*, 97:421–431.

Reddy, P. V., Ramanjaneyulu, P., Reddy, B. V., and Rao, P. S. (2020). Simultaneous optimization of drilling responses using GRA on Al-6063/TiC composite. *SN Applied Sciences*, 2(3):1–10.

Santos, R., Silva, F., Nascimento, R., Souza, J., Motta, F., Carvalho, O., and Henriques, B. (2016). Shear bond strength of veneering porcelain to zirconia: Effect of surface treatment by cnc-milling and composite layer deposition on zirconia. *Journal of the Mechanical Behavior of Biomedical Materials*, 60:547–556.

Sheikhzadeh, M. and Sanjabi, S. (2012). Structural characterization of stainless steel/TiC nanocomposites produced by high-energy ball-milling method at different milling times. *Materials & Design*, 39:366–372.

Shihab, S. K., Gattmah, J., and Kadhim, H. M. (2020). Experimental investigation of surface integrity and multi-objective optimization of end milling for hybrid al7075 matrix composites. *Silicon*: 1–17.

Shridhar, T., Krishnamurthy, L., and Sridhara, B. (2014). Machinability studies on aluminium matrix hybrid composites. *Advanced Materials Research*, 894:27–31. Trans Tech Publ.

Singh, K. K. and Kumar, D. (2018). Experimental investigation and modelling of drilling on multi-wall carbon nanotube–embedded epoxy/glass fabric polymeric nanocomposites. *Proceedings of the Institution of Mechanical Engineers, Part B: Journal of Engineering Manufacture*, 232(11):1943–1959.

Starost, K. and Njuguna, J. (2014). A review on the effect of mechanical drilling on polymer nanocomposites. In *IOP Conference Series: Materials Science and Engineering*, 64:012031. IOP Publishing.

Suresh, P., Marimuthu, K., Ranganathan, S., and Rajmohan, T. (2014). Optimization of machining parameters in turning of Al-SiC-Gr hybrid metal matrix composites using grey-fuzzy algorithm. *Transactions of Nonferrous Metals Society of China*, 24(9):2805–2814.

Swain, P. K., Mohapatra, K. D., Das, R., Sahoo, A. K., and Panda, A. (2020). Experimental investigation into characterization and machining of Al+ SiCp nanocomposites using coated carbide tool. *Mechanics & Industry*, 21(3):307.

Tabandeh-Khorshid, M., Ferguson, J., Schultz, B. F., Kim, C.-S., Cho, K., and Rohatgi, P. K. (2016). Strengthening mechanisms of graphene-and al2o3-reinforced aluminum nanocomposites synthesized by room temperature milling. *Materials & Design*, 92:79–87.

Thirumalai Kumaran, S. and Uthayakumar, M. (2014). Investigation on the machining studies of AA6351-SiC-B4C hybrid metal matrix composites. *International Journal of Machining and Machinability of Materials 2*, 15(3–4):174–185.

Verma, R. K., Singh, V. K., Singh, D., and Kharwar, P. K. (2021). Experimental investigation on surface roughness and circularity error during drilling of polymer nanocomposites. *Materials Today: Proceedings*, 44: 2501–2506.

Wagih, A., Fathy, A., and Kabeel, A. (2018). Optimum milling parameters for production of highly uniform metal-matrix nanocomposites with improved mechanical properties. *Advanced Powder Technology*, 29(10):2527–2537.

Wang, T., Xie, L., Wang, X., Jiao, L., Shen, J., Xu, H., and Nie, F. (2013). Surface integrity of high-speed milling of al/sic/65p aluminum matrix composites. *Procedia Cirp*, 8:475–480.

Xiong, Y., Wang, W., Jiang, R., and Lin, K. (2018). Analytical model of workpiece temperature in end milling in-situ tib2/7050al metal matrix composites. *International Journal of Mechanical Sciences*, 149:285–297.

Xiong, Y., Wang, W., Jiang, R., Lin, K., and Song, G. (2016). Surface integrity of milling in-situ tib2 particle reinforced al matrix composites. *International Journal of Refractory Metals and Hard Materials*, 54:407–416.

Xiong, Y., Wenhu, W., Yaoyao, S., Jiang, R., Chenwei, S., Xiaofen, L., and Kunyang, L. (2021). Investigation on surface roughness, residual stress and fatigue property of milling in-situ tib2/7050al metal matrix composites. *Chinese Journal of Aeronautics*, 34(4): 451–464.

Xu, J., Li, C., Chen, M., El Mansori, M., and Ren, F. (2019). An investigation of drilling high-strength CFRP composites using specialized drills. *The International Journal of Advanced Manufacturing Technology*, 103(9):3425–3442.

Zhong, Z. and Hung, N. P. (2002). Grinding of alumina/aluminum composites. *Journal of Materials Processing Technology*, 123(1):13–17.

Zhu, Y. and Kishawy, H. (2005). Influence of alumina particles on the mechanics of machining metal matrix composites. *International Journal of Machine Tools and Manufacture*, 45(4–5):389–398.

Zinati, R. F. and Razfar, M. (2014). Experimental and modeling investigation of surface roughness in end-milling of polyamide 6/multi-walled carbon nano-tube composite. *The International Journal of Advanced Manufacturing Technology*, 75(5–8):979–989.

3 Advanced Machining of Nanocomposites

CONTENTS

3.1 INTRODUCTION

Nanocomposites have become increasingly popular in modern industries due to their superior mechanical characteristics, high strength-to-weight ratio, high wear resistance, and exceptional corrosion resistance. Their applications, including in aviation, shipping, and transportation industries are increasingly looking for low-density materials that are strong and rigid and abrasion, impact, and corrosion resistant. Many modern techniques require materials with fairly unusual combinations of properties that are generally quite rare in traditional materials. In that regard, there is always a need for materials that possess these properties; for such applications the advancement (Callister and Rethwisch 2018) of nanocomposites has emerged as a utilitarian material (Singh and Agarwal 2016). The essential criterion for these products is that they be economically processed. Conventional machining or powder metallurgical process will be appropriate under simplified conditions for basic geometries. But they also require extensive post machine operations, which are mostly incapable of manufacturing complex geometrical features like deep cavities or small inner radii, transverse openings, undercuts, bevels, slots, and threads, besides some more complicated parts with tighter tolerances requiring machining, making the products more costly. Particularly, the manufacturing of tailor-made, complex, engineered nanocomposite parts in smaller batch sizes is a significant economic concern if traditional final machining technologies are used.

Therefore, there is a need for the evolution of improved cutting tool materials in machining, so that productivity is not affected. Materials such as titanium; nimonics; and other similar high-strength, temperature-resistant alloys; fiber-reinforced composites; stellites (cobalt-based alloys); ceramics; and difficult-to-machine alloys

DOI: 10.1201/9781003107743-3

are required for machining such stronger and harder materials (Jain 2009). It is more difficult to work with intricate and complex shapes on such materials. Other high expectations necessitate higher production rates, low tolerance values, complex structures, automated data transmission, miniaturization, etc. (Jain 2009; Snoeys et al. 1986). Producing holes (shallow entrance angles, noncircular, microsized, large aspect ratio, large no of small holes in one workpiece, contoured holes, hole without burrs, etc.) in nanocomposites is another area where relevant processes are necessary. To meet these requirements, a separate class of machining processes have been evolved.

3.2 ADVANCED MACHINING

Manufacturing is considered as the basis of any industrial society, and the level of manufacturing activity defines a nation's economic status. To improve productivity and diminish the cost of the product, manufacturing systems' efficiency has been improved and innovative manufacturing technologies, automatic systems, and various innovative systems have been implemented during the last few decades. The consecutive developments of machine tools, digital control and intelligent level with greater speed, high precision and flexibility followed. With the invention of sensors, incorporation of probes, and adaptive and hybrid manufacturing, the capability of machining operations has been enhanced to a great extent. The sophisticated knowledge base, virtual reality, and internet of things make the workstation truly intelligent. Mass customization in manufacturing also inspired the adaptation of advanced concepts that made the manufacturing operation more flexible, adaptable, and sustainable.

3.2.1 Electrical Discharge Machining

Electrical discharge machining (EDM) is amongst the earliest advanced machining processes wherein the material is got rid of by a precisely controlled spark generated between two electrodes in the presence of a dielectric medium. It is one of the most widely used and widely beneficial methods in the injection mold and die-making industries to generate intricate shapes and mold cavities. The distinctive feature of using thermal energy to machine conductive materials of almost any hardness has proven useful in the production of molds, dies, automotive as well as aeronautic components, electronics, home appliances, machineries, packaging, telecommunications, accessories, toys, and surgical instruments (Pradhan 2010; Jahan 2019). Figure 3.1 illustrates the basic components of the EDM process.

Every discharge melts and evaporates material from the workpiece and electrode under normal operating conditions. Multiple discharges create a typical surface pattern (Figure 3.2, similar to the moon's landscape. This is common in metal EDM machining. Some molten material can resolidify on the surface, forming a distinctive white layer.

Research has been performed and authored on the machining aspect of MMCs with particle reinforcement. In the late 1990s there have been a few publications related to EDM of ceramic composites, a sample made of Al2O3 + TiC (Lok and Lee

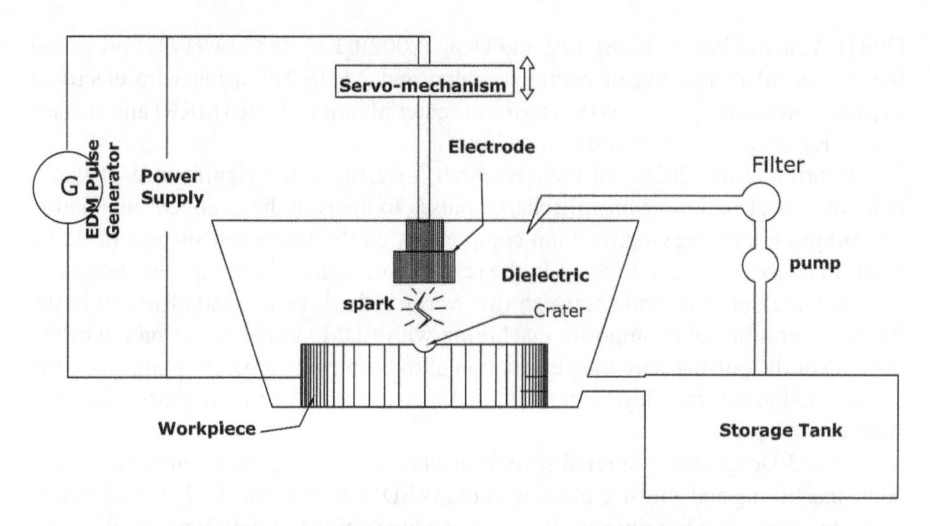

FIGURE 3.1 Basic components of the EDM process. (From Pradhan 2010.)

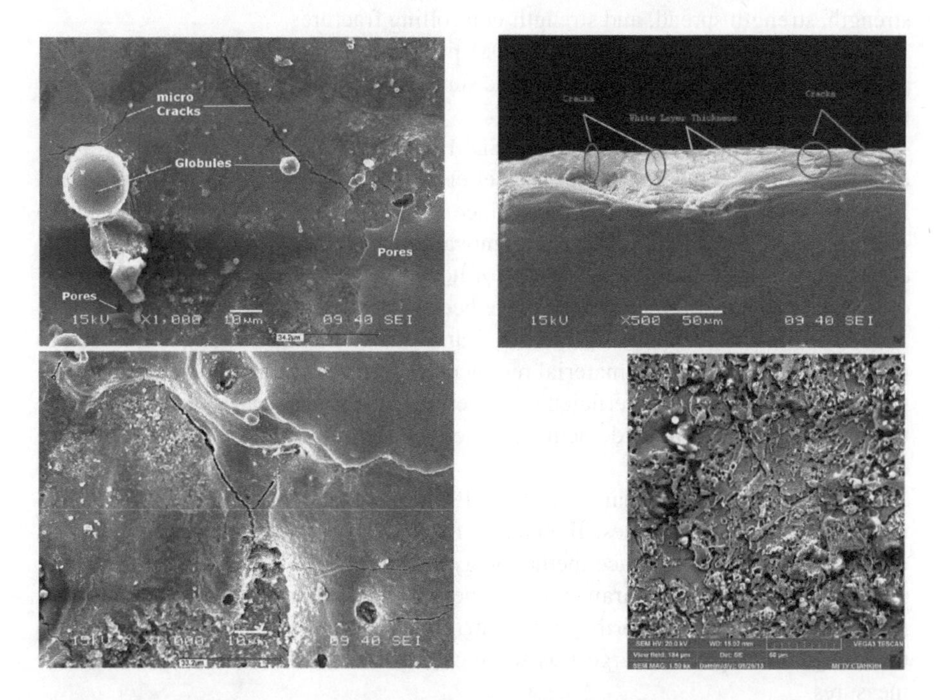

FIGURE 3.2 SEM image, showing the surface condition, white layer, cracks, etc., after the electrical discharge machining of AISI D2 and the chrome workpieces. (From Pradhan 2010, 2013; Pradhan and Biswas 2010; Volosova et al. 2015.)

(1997); Yun and Wang (1998); Lee and Deng (2002)). Lok and Lee (1997) presented the successful machining of ceramics sialon and AI_20_3–TiC using wire electrical discharge machining (WEDM) in terms of material removal rate (MRR) and surface finish (Ra) under different cutting conditions.

Grigoriev et al. (2020), in their research, investigate the significance of nanocomposite destruction against discharge pulses to analyze the chemical interaction of workpiece and working medium components on the machined surface made by electron microscopy, and to identify the resulting ceramic nanocomposite defects.

A comparison was made between two working fluids (water and olive oil) in the EDM of ceramic nanocomposite machining with EDM variables, evaluation of the conduct of the pulsing wire tool, evaluation of the content and medium impact of the surface and subsurface layers method, and measurement of the discharge distance were carried out.

Lee and Deng (2002) studied several mechanical surface treatments, viz., ultrasonic machining and abrasive blasting of the WEDMed surfaces of AI_20_3/TiC/Mo/Ni ceramic composites to minimize the surface contribution to fracture probability, and characterize dependability alteration and machining damage as a result of flexural strength, strength spread, and strength-controlling fractures.

Lee and Deng recommended for EDM pulse generator parameters, although an acceptable characteristic of the machined surface that could be suitable for industrial use was not achieved.

The machinability of ceramics by EDM has been taken into account by a few authors and their research groups (Patel et al. 2009, 2010, 2011), who optimized EDM parameters and analyzed the surface after EDM. Efforts have been made to investigate the surface integrity and removal mechanisms of the AI_20_3–SiCw–TiC ceramic composite EDM. The impact of normal, as well as rough cutting modes of surface and subsurface damage, have been considered. Observations of surface and subsurface impacts, as well as EDX analysis of EDM surfaces, were used to predict the mechanisms of material removal. Material removal may be used advantageously to enhance the efficiency of ceramic composite parts. In view of the thermal self-induced flake detachment, the material removal rate could be greatly enhanced.

EDM was used by Mahanta et al. (2019) to explore the Al7075-B_4C fly hybrid metal matrix nanocomposites. To enhance machinability, the Taguchi approach was utilized, and response surface methodology (RSM) was used to assess the connection between machining parameters and performance metrics. The L18 Full factorial design with a mixed orthogonal matrix was used for the trials. According to the study's findings, discharge was the most significant factor in the performance measures.

Melk et al. (2016) investigated the machinability of 3Y-TZP reinforcement CNT composites using EDM. The damage was investigated and the mechanisms for removing the material were determined. It has been found that the inclusion of CNTs in the zirconia matrix causes a strong increase in electrical conductivity and a relatively minor change in thermal conductivity. George et al. (2004) studied carbon–carbon composites to assess the machinability and the effect of machining variables on the response, considering three parameters at two stages, and found that pulse current and duration pulse are important for electrode wear rate (EWR) and MRR. The influence of SiC percentage volume and other processing characteristics was investigated when Al–SiC was machined, and the improvement in SiC reduces MRR when EWR and surface roughness (SR) significantly improve (Singh et al. 2004).

The influence of electrode spinning on the machining of Al–SiC and Al–$AI_2 0_3$ composites have an immense effect on the MRR, EWR, and SR (Mohan et al. 2002). Gopalakannan et al. (2012) worked with the EDM characteristics of Al 7075/10wt% of SiC composites in determining the machinability and quality of the workpiece, and the results revealed that the MRR exhibited an increasing pattern with an improvement in the product of pulse current and pulse-on time has been reached to its maximum value and then reduced.

Gopalakannan and Senthilvelan (2013) presented an experimental assessment on the machining parameters for the powder-mixed EDM of Al–10% of SiC MMC. They combined silicone powder in the dielectric liquid and confirmed that the application of silicone powder to the dielectric liquid of EDM improved MRR and decreased SR. The EDM of nanocomposites associated magnesium nano-aluminum composites developed by a powder metallurgy was evaluated and confirmed that the pulse-on time showed a significant effect (Ponappa et al. 2010). In general, the literature on the machining of metal matrix nanocomposites (MMNCs) is limited, and contains investigations mostly on the characteristics of EDM.

The study of Patel et al. (2011) explores the effects of EDM of Al_2O_3–SiCw–TiC ceramic composite. The impacts of process parameters on the MRR, EWR, and Ra were carried out by conducting experimentation on the die lowering EDM machine in accordance with the planned central composite rotatable design. Three different empirical second-order models were created and verified for each response and claimed to be reasonably good predictors of the regression model. In addition, SEM was used to evaluate and describe surface or even subsurface damage.

Besides that, the influence of standard and rough cutting modes on surface and subsurface damages were discussed. Surface and subsurface damage assessments, as well as EDX analyses of EDMed surfaces, have been utilized to anticipate material removal mechanisms. According to the existing references, Al_2O_3+SiC+TiC composites and nanocomposites produced by manufacturing techniques, including 30%–40% reinforcements, are electrically appropriate for EDM. Conductivity in such composites has often been substantially higher than the conductivity limit for products that may be considered to be conductors. Several studies have shown that even nonconductive ceramic composites with more than 40% conductive addition may be ideal for EDM. Conversely, the desired precision of the machined surface cannot be achieved. Liu and Huang (2003) studied the effect of the EDM operation on the surface integrity, machinability, and impact resistance of hot-pressed conductive Si_3N_4-TiN composites. Severe microdamage and greater surface roughness have been observed due to high pulse energy, i.e., higher working voltage and current. Pradhan and Dehari (2019) conducted an experimental study in which pulse current (Ip), pulse duration (Ton), V, and Tau were chosen as process parameters affecting MRR, TWR (tool wear rate), Ra, and Radial Overcut (ROC) of Al7075+3%B_4C+7%TiC hybrid composite. Investigators found that Ip is the most effective in terms of EDM machining characteristics when compared to voltage.

Gopalakannan and Senthilvelan (2013) attempted to explore the physical properties of MMNC and EDM experiments by using RSM to establish a statistical equation and examine the effect of the parameters on MMNC system output. The numerical mathematical equations were thus developed to explore the effect of input parameters on the responses using RSM. Zeller et al. (2017) studied the EDM efficiency of SiC/GNP nanocomposites with distinct GNP content (10 and 20 vol.%) by contrasting respective reactions to three related SiC monolith materials, to establish relationships between the reaction of the EDM and the transport properties of the materials. In addition, machining studies have been performed under different energy levels, and graphene-related nanocomposites were first evaluated on orthogonal surfaces based on their anisotropic microstructures.

Gommeringer et al. (2020) prepared composites with a yttria neodymium co-stabilized TZP matrix and electrically conductive niobium carbide dispersion by hot-pressing, and investigated the effect of changes in parameters on the EDM characteristics that are crucial for effective technology implementation.

In addition, an enlarged SEM image of the range of workpieces after electric discharge machining of the chrome workpiece is presented (Figure 3.2). The image shows holes, cracks, drops of hardened molten pool as a result of impacting the discharge channel during machining.

Nanocomposites are 21st-century materials, with a growth rate of 25% due to their multifunctional capabilities. With distinctive design potential and attributes, they draw recognition amongst researchers across the world (Rathod et al. 2017). Nanocomposites are expanding their capabilities in several advanced applications such as aerospace, biotechnology, and many others in future missions, owing to the possibility of combining the desired properties by distinct reinforcement. The preference for nanocomponents (matrix and nanofillers) leads to the addition of some preferred properties. The properties of nanocomposites with different reinforcement like ceramics and others, with hardness; resistance to abrasion, corrosion, and erosion; and many other superior mechanical and wear properties make them suitable for many advanced applications (Tavangarian et al. (2018); Grigoriev et al. (2019); Hu et al. (2018); Mathiazhagan and Anup (2018); Wu et al. (2018)).

As for machinability, the uses of MMCs are constrained due to their low machinability, which results in low surface finish and excessive tool wear. Thanks to the higher stiffness and strength of the reinforcement, it is difficult to process MMCs using conventional techniques. Therefore, the EDM cycle has become a feasible approach for these forms of MMCs. The EDM technology does not necessitate mechanical energy, and the material removal rate is unaffected by material properties such as hardness, strength, toughness, etc. Products with low machinability, like cemented tungsten carbide and also composites, could be manufactured relatively easily using the EDM process (Gopalakannan and Senthilvelan 2013).

Quarto et al. (2019) present an investigation into the suitability of the ZrB_2-reinforced SiC fiber EDM process. Circular pocket features have been machined using a micro-EDM machine. Aspects of performance characteristics were computed based on a single discharge number and energy. The results show interesting aspects of the process from the point of view of both productivity and removal mechanism.

Pradhan and Tiwari (2017) and Tiwari and Pradhan (2017) have seen the effect of Rice Husk Ash (RHA) particle wt% on mechanical properties, in addition to the study of the machinability of aluminum alloy LM25 using EDM to see the impact of RHA particle wt% and other machining parameters like MRR, TWR, and surface roughness. Furthermore, Ip is a significant parameter for MRR, which indicates a growing pattern with Ip, as seen with TWR, while Ra rises with differing percentages of RHA ranging from 4% to 12%. Mausam et al. (2019) applied gray relation analysis to establish the appropriate parameters, such as pulse-on time, gap voltage, and peak current, as well as their important impacts on MRR and TWR. Lastly, the values chosen using the gray analysis resulted in a substantial boost in MRR and a decrease in TWR.

EDM is used primarily for conductive materials and, therefore, the mechanical, physical, and metallurgical characteristics of the material's surface do not impact the properties of the work material significantly. EDM has long been the solution for high-precision, complex machining applications where traditional metal removal is challenging or impossible. It can perform various types of operations, viz., ED drilling, ED milling, 3D shaping by wire-cut EDM, and many such hybrid ED grinding methods. Its significant characteristics are indeed predictable, consistent, and repeatable operations. If one could see modern developments over the last decades or so, one might realize how a lot of significant changes have taken place. The hybridization of various technologies, such as artificial intelligence and machine learning, as well as the Industrial Internet of Things (IIoT), also referred to as Industry 4.0 or the Fourth Technological Revolution, will always change the manufacturing sector from the machine tool industry to quality assurance.

EDM is an excellent companion to create tools, dies, or related products. EDM has the potential to create intricate and irregular forms from extremely hard materials. Previously, EDM machining was mostly used in machine tool applications; however, with technology developments, it became commonly included in cutting highly complicated forms, including automobile and aircraft/spacecraft parts.

Many researchers have successfully carried out EDM of nanocomposites. Machining conditions with modest thermal energy have a tendency to appropriate machining speed, low surface roughness, and almost defect-free surfaces of very thin areas affected by the machining operation. Under such conditions, the dominant mechanism for the metal removal process is mostly melting only. Detailed adequate machining conditions and adjustment of composite composition are needed to further improve the productivity and efficiency of the machining process. Findings obtained using various algorithms have been reported by the researchers. Remedies are compared to several contemporary algorithms, such as the Firefly Algorithm. Performance is compared to the RSM approach, and further hybridization with other tools, viz., MCDM, is reported. A comparative analysis is also reported.

The effect of various machining parameters along with different reinforcement compositions and their wt.% has a significant role in machining performance and surface quality. Pradhan and Singh (2019) did, however, investigate the influence of those parameters on EDMed Al7075–SiC–WS_2 hybrid composite and multi-objective variable decision-making parameters. Pradhan (2020) also investigated the effect of composition with machine parameters using the Taguchi L9 orthogonal array for

AA-2014 hybrid composite reinforced with SiC and glass particulates and found the composition and Ip to be the most influencing parameters for machinability.

This chapter presents an overview of the challenges of EDM processing of nanocomposite materials. When compositions of these materials were machined, the parameters associated with the machining and the EDM behavior were reported. Several methods for material removal have been reported, including melting, degradation, and oxidation/decomposition. The dominant parameters were the material composition, material, microstructure, and EDM parameters. This means that in order for technology to be developed, the procedure correlation must be thoroughly understood. EDM, on the other hand, can be used to effectively manufacture various nanocomposite components.

3.2.2 Electrochemical Machining

Electrochemical machining (ECM) is regarded as one of the modern machining methods available today for machining nanocomposites. Faraday's electrolysis principles provide the basis for ECM. It utilizes electrical energy in combination with chemical reactions by anodic dissolution to accomplish material removal, which is a reverse process of electroplating, known as deplating. The required geometry of the parent material is obtained through anodic dissolution and is specific to a wide range of conductive materials. The workpiece shape must be created with the electrode tool, wherein, the tool is the cathode and the workpiece is the anode. With the development of many applications such as turbine blades, aeroengines, etc., the ECM technology has consistently shown some progress in processing efficiency, speed, and cost in recent decades (Xu and Wang 2019), Soni and Thomas 2017). ECM has been regarded important to many other types of processing, particularly when working with "exotic" materials where fractures must constantly be avoided and a higher surface quality is required, such as in the aviation industry.

Figure 3.3 shows a schematic illustration of a typical ECM process. The positive power supply connection is attached to the workpiece and the negative terminal to the tool. ECM is among the most effective techniques for machining hard materials with complicated shapes. Hard materials machined by ECM exhibit superior surface integrity as well as precision, making them one of the most preferred and widely used in sectors such as the aerospace industry, for manufacture turbine blades, for the production of dies, and so on. As a consequence, ECM process capabilities continuously need to be improved (Figure 3.4).

Because ECM is a contactless machining technique, it is best suited for difficult-to-cut materials, irrespective of their hardness and strength. Since only hydrogen is emitted on the cathode, the formed cathode tool does not wear during the process. Besides this, high material removal speeds, high machining precision, and reasonable surface quality could be attained without the risk of deformation, microcracks, residual tension, recast coating, or heat-influenced areas.

ECM processing capabilities may be employed in various operations in several different applications, including the aviation sector, where they can be utilized to make high aspect ratio straight, slanted, and turbulated cooling openings for turbine cutting edges, useful gaps in aviation segments; streamline sealing; in the vehicles

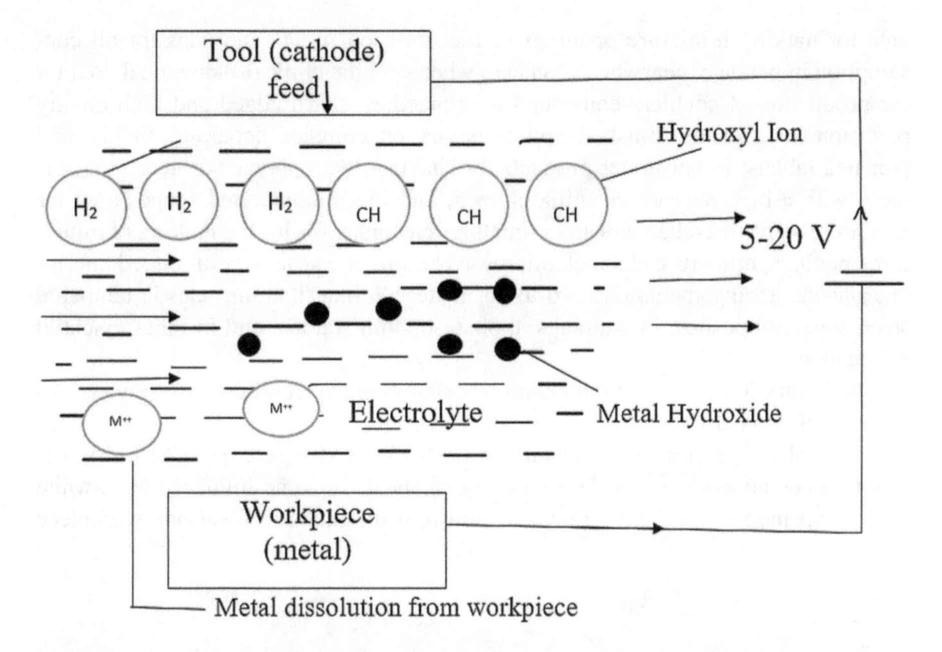

FIGURE 3.3 Principle of electrolysis.

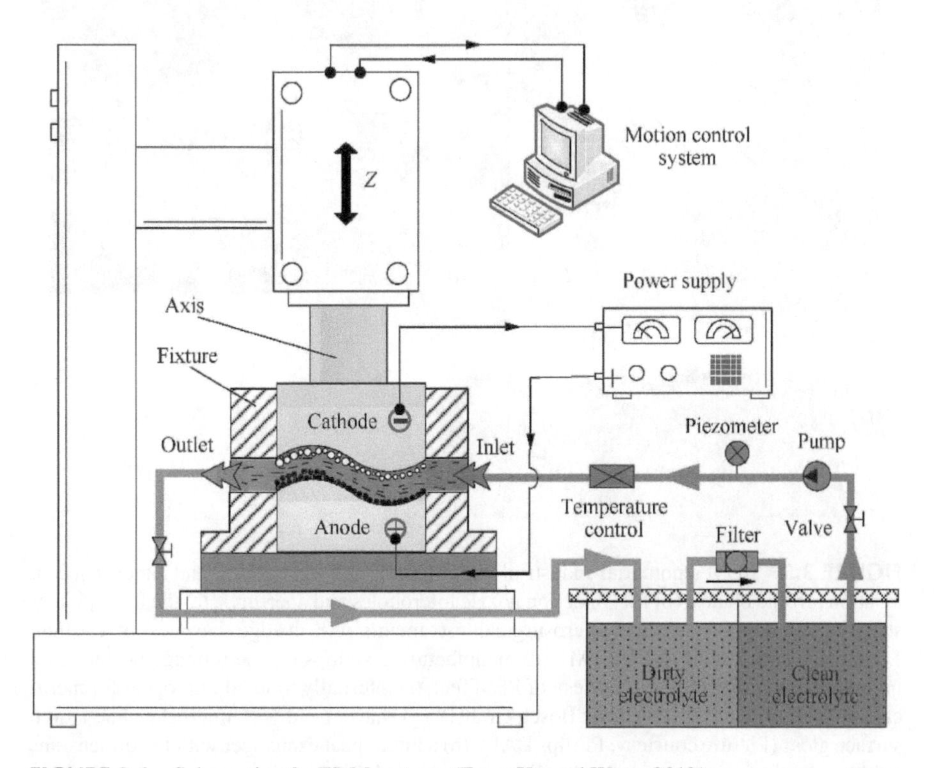

FIGURE 3.4 Schematic of a EDM system. (From Xu and Wang 2019.)

area for making miniature openings in fuel infusion spouts and gaps for oil consumption in passage, gearwheels, and cogwheels; in the clinical/biomedical area for the production of cochlear-embedded microneedles, sharp-edged and high-quality precision instruments, finished optical passes on complex apparatus molds, and punched tablets; in consumer products for shaving, hydrophobic surfaces, and surfaces with a high surface cleanliness need, and in the compound shapes area for miniature warmth exchangers and miniature reactors, tools for the making of miniature openings, miniature channel, microwave, miniature spaces, confounded interior and outside arrangements on hard-to-cut materials like titanium, elastic tempered steel, supercomposites, instruments, tool stash, coin stamps, and in other essential applications.

The Figure 3.5 gives a graphical representation of several industrial applications wherein ECM is utilized.

An SEM image of an ECMed surface can be seen in Figure 3.6. The cross section reveales no evidence of thermal or mechanical rim zone influence (e.g., white layer, deformed grains, etc.). Local nonuniform dissolution of various workpiece

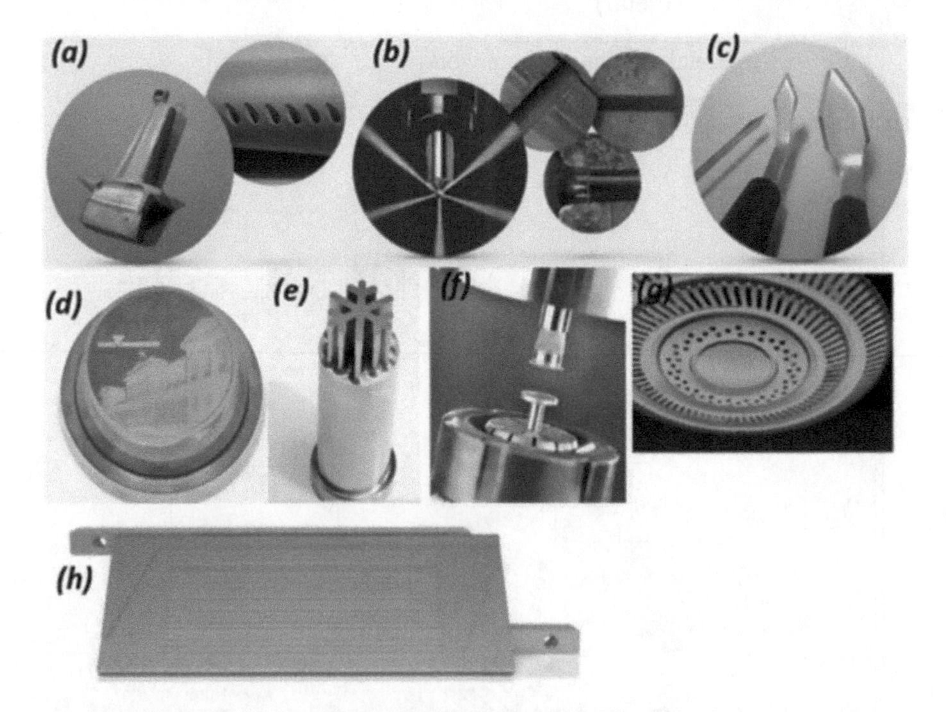

FIGURE 3.5 EDM's potential industrial uses: (a) aperture for cooling and functioning in aviation turbine blades, (b) fuel injection nozzle microholes and apertures, (c) the formation of sharp cutting edges on medical microsurgical instruments with an edge sharpness of less than 1 m (Image courtesy of INDECTM), (d) manufacturing stamps for the mintage business; (e) micro punching dies (Image courtesy of PEMTec), (f) internally toothed microgear for medical applications (Picture courtesy: Bosch GmbH), (g) shaver head with microslots and a high surface gloss (Picture courtesy: Philips DAP), (h) a micro heat exchanger with the dimensions (width x depth) for heating channels. (From Saxena et al. 2018.)

FIGURE 3.6 Microscopic images of material alterations during direct current ECM: flow grooves and local dissolution. (From Klink 2016.)

material (micro)phases may occur depending on machining conditions, resulting in process-induced roughness or waviness. Local corrosion resistance and the formation of oxide layers must be regarded as process-dependent material changes in this context.

Although ECM is a multi-field, coupled process with numerous influencing elements, machining precision and machining stability are challenging to manage. ECM has significant drawbacks, particularly for the tool electrode design necessary for complex profile processing, which frequently necessitates repeated adjustments.

There is room to improve ECM in three significant areas. The objective is to improve electrolyte stream conditions in the between terminal hole, upgrade anodic disintegration limitation, and look after little, stable holes to accomplish higher machining exactness, great surface quality, and better interaction steadiness. Advances utilized in the airplane industry incorporate ECM and electrochemical release machining. Electrochemical mixture machining is another field wherein electrochemical energy structures are consolidated to create parts while different cycles are running.

Yadav et al. (2018) investigated the kerf characteristics of a straight cut in alumina epoxy nanocomposites using the wire electrochemical spark cutting (WECSC) technique. Kerf characteristics such as kerf deviation and taper have been used to evaluate cut efficiency. The applied voltage, electrolyte concentration, wire velocity, pulse-on time, and pulse-off time too are varied, and the impact on the kerf characteristics is analyzed. The applied voltage, electrolyte concentration, and wire velocity were found to be important parameters for straight cutting of alumina epoxy nanocomposites. The WECSC process for straight-cutting alumina epoxy nanocomposite is a useful machining approach for nanocomposite materials. Kerf deviation is reduced where the applied voltage is low and the wire velocity is high.

Using the WECSC technique, Yadav et al. (2019) examined the machining activity of silicon dioxide (silica) epoxy nanocomposite for straight cutting. To investigate the influence of manipulating input parameters, a specific set of tests were carried out using the one-parameter-at-a-time method. The influence of the different machining parameters like voltage supply, electrolyte concentration, wire velocity, pulse-on time, and silica particle concentration (Cp) of 3%, 4%, and 5% (wt.%) on performance measurements like MRR and surface roughness was investigated. WECSC has been described as a valuable method for cutting silicon dioxide epoxy nanocomposites, where MRR is affected by decreasing the silica particle content.

Rajkumar et al. (2015) has modeled and optimized machining parameters in the abrasive-assisted EMC of aluminum–boron carbide nanocomposites. To approximate the surface quality and the MRR, a statistical RSM was used with four input parameters: current, voltage, feed volume, and concentration. For the same input parameters, it has been determined experimentally that abrasive-assisted ECM produces higher MRR in the case of Al–nanoB_4C composite. As a result, the decrease in reinforcement size to the nano level has allowed hardening without compromising machinability (when machined by abrasive-assisted ECM). Generally, hardening materials lose their machinability. However, this method of hardening (reduction of reinforcement size) seems to be unique and superior to others. This conclusion emphasizes the significance of this method in the manufacturing of materials with complex forms and high hardness.

Prakash and Gopalakannan (2020) investigated micro electrochemical machining (μECM) of the aluminum alloy AA-7075 reinforced with nano-silicon-carbide particles. The desirability analysis of multiple-response adaptation was carried out, an analytical equation that correlates all outputs was created, and the most important input parameters were identified using ANOVA. In practice, a teaching-learning-based optimization (TLBO) algorithm was employed to optimize the surface roughness as a constraint, in order to compare the results with a desirability analysis.

Senthilkumar et al. (2009) explored the impact of process factors including applied voltage, electrolyte concentration, electrolyte flow rate, and tool feed rate on MRR and surface roughness in LM25 Al/10%SiC stir cast composites. Contour plots are generated to examine the influence of process parameters and their interactions. RSM is used to refine the process parameters. The experimental research focuses on electrochemical machining standards such as MRR and Ra in ECM, which are heavily affected by machining parameters.

In the laboratory, Jain et al. (1991) designed, developed, and fabricated an apparatus for traveling wire electrochemical spark machining (TW-ECSM), and experiments were conducted on glass epoxy and Kevlar epoxy composites with NaOH as the electrolyte, and it was discovered to be a feasible and effective process. Voltage and electrolyte concentration were assumed to be controllable variables, and their effects on MRR, diametral overcut (O,), TWR, and wire erosion ratio (WER) were the process's responses. Lowering the voltage and electrolyte concentration results in higher machining accuracy. Thermomechanical metal removal was revealed to be the most important phenomenon in TW-ECSM. MRR increased with increasing voltage, along with the presence of thermal cracks, a large heat-affected field, and rough machined surfaces.

Liu and Yue (2012) concluded that electrochemical discharge machining (ECDM) is a more stable process than EDM. A surface finish in the submicron range can be achieved with proper equipment development and electrode design. For the highest MRR, orthogonal analysis was conducted and optimal conditions for electric current, pulse duration, and electrolyte concentrations were calculated. Each variable is divided into three categories. An L9 orthogonal design table was created to improve MRR. An orthogonal evaluation of 10ALO and 20ALO materials was performed with this goal in mind, and the results are presented. As per the findings, the most important variable at the moment is applied current, with ECDM applied to AI_20_3 particulate MMCs to achieve a high MRR between pulse length and electrolyte density.

Rao and Padmanabhan (2013, 2014) performed an electrochemical machining operation on LM6 Al/B_4Cp composites. B_4Cp particles of 30 micron size were hardened with 2.5%, 5%, and 7.5% by weight in an LM6 Al alloy matrix. The L27 orthogonal array by Taguchi was used to analyze different machining variables. To validate the results, tests with optimum amounts of machining parameters were performed. The experimental results demonstrate that it can increase the responses in ECM. During the experiment, LM6 Al/B_4C composites were machined using an ECM technique, and the machining parameters were optimized using Taguchi's system. Taguchi orthogonal arrays, signal-to-noise (S/N) ratio, ANOVA, and regression analysis are used to obtain optimal levels as well as realize the effects of different process parameters on MRR, surface roughness, and radial overcut values.

Attention is needed in the development of machine tools for practical use, micron level electrochemical dissolution and solutions to locate material removal, evaluation of machinability of various materials, evaluation of process capability, effects of process parameters, development of improved instruments, electrolyte selection, and ultrashort pulsed power development supplies. A great deal of work is required in developing small sources of pulsed energy that are cheap and reliable. There is very little literature that correlates ECM pulses with the mechanical performance required to monitor them in the process. Several techniques must be devised to prevent the workpiece from being immersed in the electrolyte. Micro-ECM is a multidisciplinary process that involves engineers, chemical engineers, physicists, and chemists. Shape memory alloys, cermets, additively fabricated materials, sandwich materials, semiconductors, and other novel materials have yet to be adequately investigated in terms of machinability. More research is needed on dissolution processes for tool materials such as WC-Co with various percentages of Co- and SiC-based ceramics.

Various research steps have been taken related to the machining of components for various applications, such as electrochemical dissolution properties of modern difficult-to-cut materials, computational simulation of electrochemical operation, design for complicated profiles and configuration of cathode tool, flow field simulation, design for uniform electrolyte flow, and development . The recent ECM trends for the fabrication of aeroengine components have been explored. Dissolution properties, machining precision, machine tools, new processes, and intelligent ECM are the five major study fields. With more study, ECM can be widely used in the fields of aeronautics and astronautics.

3.2.3 Laser Beam Machining

Laser beam machining (LBM) is a popular machining technology that employs a laser beam as a heat source. The laser beam is focused to melt and vaporize the undesirable material from the parent material. It is used to shape nearly all engineering materials, including composites and nanocomposites. Because of LBM's ability to machine a variety of different materials at high precision, it has become a popular machining method. It is generally used for cutting, boring, checking, welding, sintering, and heat treatment. The laser is frequently used in turning and milling operations, but its primary application is in the cutting of metallic and nonmetallic sheets. It can also be used as a support tool for conventional machining, i.e., turning, to improve machinability by modifying the microstructure and softening the surface of the part. By heating the part before the final operation, the laser beams reduce machining strength and improve the cutting characteristics. Laser-based processes such as laser re-melting, laser hardening, laser cladding, and laser alloying have been extensively used in surface engineering applications. LBM has numerous applications in the automotive, aircraft, electronic, civil, and nuclear industries, as well as in household appliances. Stainless steel, a distinctive engineering material used in automobiles and household appliances, is ideal for laser beam cutting (Ghany and Newishy 2005; Yilbas et al. 1992).

Melting, evaporation, and chemical degradation – including chemical bonding and material degradation – are the material removal mechanisms in LBM. The thermal energy needed for fusion and evaporation is attained by absorbing a high-density laser beam focused on the workpiece via the glass. For clean cuts with sharp edges, a high-pressure assist gas may be used to extract melted, vaporized, or chemically eroded materials from the machining area.

In a range of machining applications, laser hybrid machining methods have been proven to be superior to a single nonconventional machining approach. LBM is a noncontact advanced machining technology with considerable flexibility, but because of the thermal nature of the process, careful management of the laser beam is required to minimize any unwanted thermal effects. 3D LBM procedures are not completely developed, and much more study is needed until these could be used in industry.

Biswas et al. (2010) experimented with pulsed Nd:YAG laser microdrilling on titanium nitride alumina composite. The RSM was used to investigate the influence of the five laser microdrilling process parameters, namely, lamp current, pulse duration, pulse distance, assist air pressure, and focal length. When the focal length was centered on the material surface, it was considered zero, and when it was set above or below the surface, it was considered positive and negative, respectively. A mathematical approach is used to test and model three geometrical features: the circularity of the hole at the entrance, the circularity of the hole at exit, and the hole taper. ANOVA was used to determine the process's major variables.

Roy et al. (2015) studied Nd:YAG laser microdrilling of SiC-30BN nanocomposite material which, due to its machinability, is not suitable for milling using traditional machining processes. These needs led to the creation of innovative machining methods, called nontraditional machining processes (NTM). Neodymium (Nd) atoms are incorporated into a yttrium aluminum garnet (YAG) crystal host in Nd:YAG lasers. According to the findings of this research, the laser beam microdrilling technology may be utilized for effectively microdrilling SiC-30BN nanocomposites. The laser microdrilling of SiC-30BN nanocomposite with numerous performance analysis is optimized using Taguchi-based gray relational analysis (Figure 3.7).

A detailed schematic of LBM is presented in Figure 3.8. LBM's ability to cut complicated forms and drill microsized holes with near tolerances in a wide range of materials has opened up new opportunities for industries. LBM micromachines coronary stents utilized in the medical industry. In comparison to other thermal-energy-based techniques like EDM and ECM, it has a lower heat-affected zone (HAZ), making it ideal for micromachining applications. Figure 3.9a and b shows a complex form for medical purpose in a metal coronary stent cut by pulsed Nd:YAG laser. As illustrated in Figure 3.9c and d, laser beam micromachines YAG laser beam cutting and glasses are used in optoelectronics.

The use of carbon-fiber-reinforced polymer (CFRP) composite materials as a high-performance lightweight in the aerospace and automotive industries necessitates the development of efficient and high-performance machining technologies (Negarestani et al. (2010)).

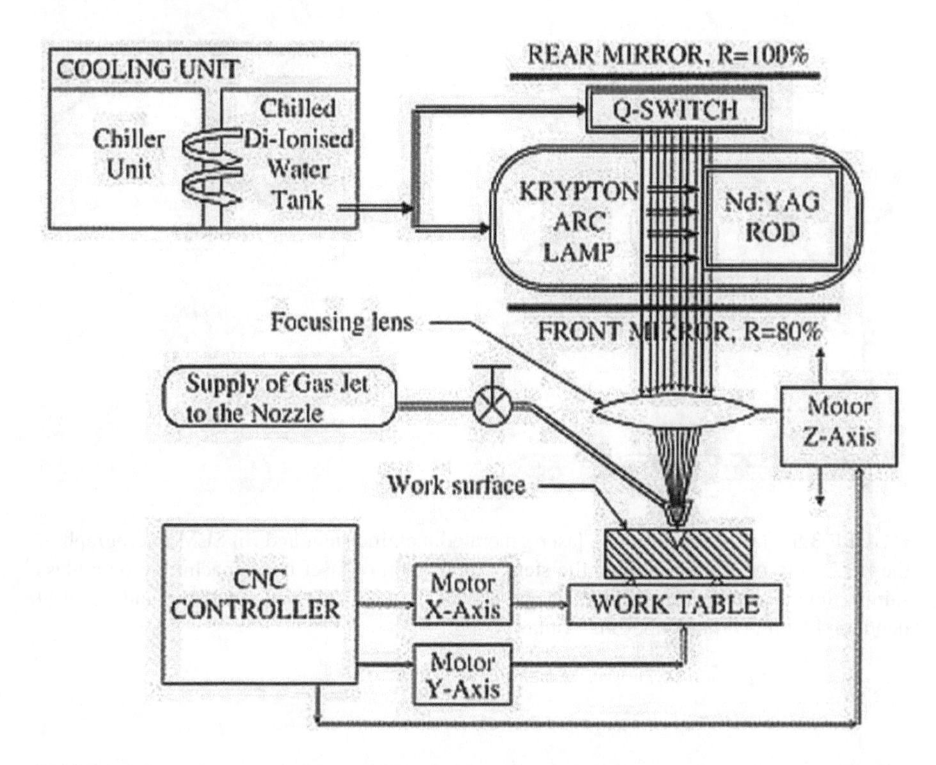

FIGURE 3.7 Schematic representation of a laser machining system. (From Kuar et al. 2006.)

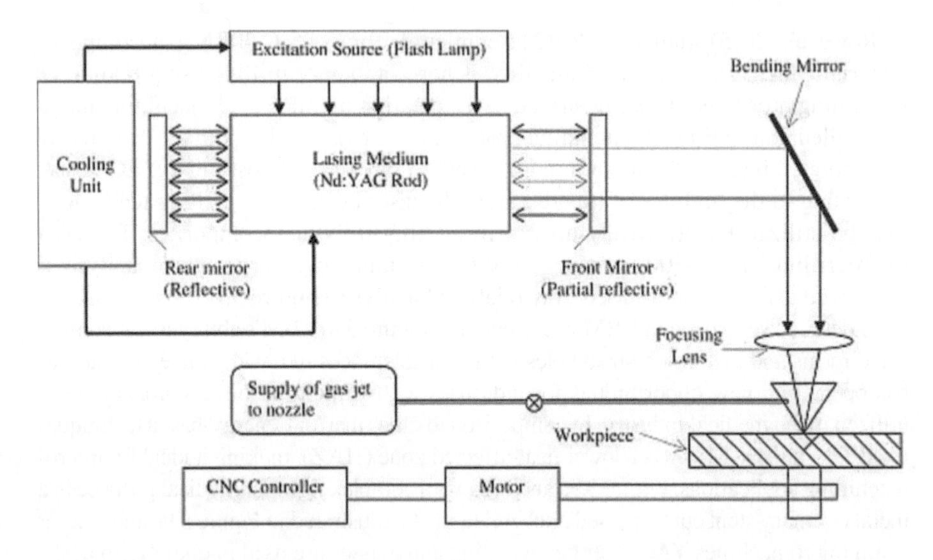

FIGURE 3.8 Schematic of a Nd:YAG laser beam cutting system. (From Dubey and Yadava 2008.)

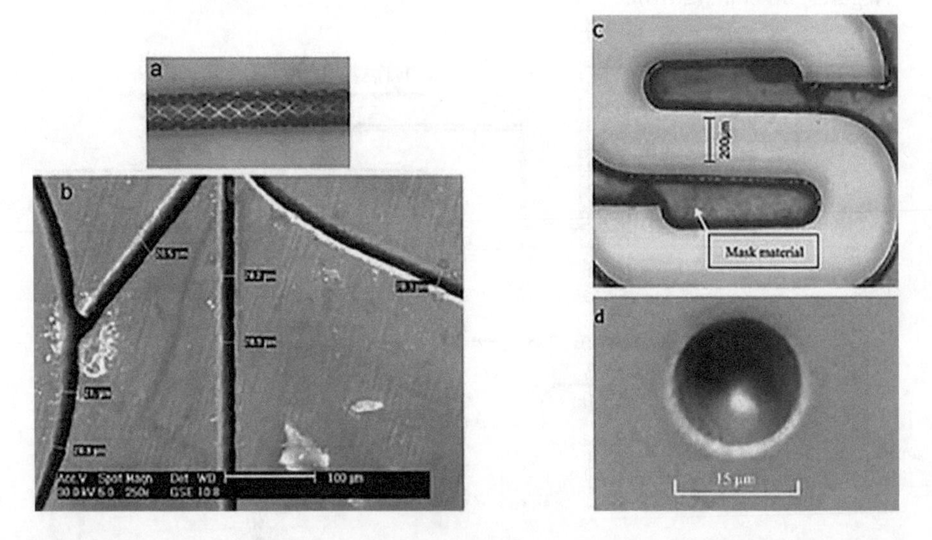

FIGURE 3.9 (a) Illustration of a laser-generated metallic stent and (b) SEM micrograph of the kerf width of a laser-cut metallic stent. Application of laser micromachining on a glass sample: (c) intricate features and (d) a spherical cavity with a diameter of 15 µm and a central depth of 4.5 µm. (From Dubey and Yadava 2008.)

The large thermal property difference between the two constituent materials (60% carbon fiber–40% epoxy polymer) results in a significant heat-affected zone. They developed a 3D FE model of heat flow and material removal during ultraviolet laser machining of CFRP composites that has been using a formulation of the process conditions and material properties. The results of FEA (finite element analysis) simulations showed similar qualitative and physical agreement with the experiments. They predicted HAZ and ablation depth to be more sensitive to speeds in the lower range of 50–200 mm/s compared with the higher range of 200–800 mm/s. The feasibility of using FEA modeling for simple (single-line cutting) and more complicated (double-line cutting) machining of CFRP composites using heterogeneous meshing at scanning speeds greater than 200 mm/s has been demonstrated.

Negarestani and Li (2012) experienced machining challenges particularly in the case of CFRPs. This is due to the large difference between mechanical and thermal properties of the constituents. Therefore, alternative machining processes are being studied, including abrasive water jet machining (AWJM), laser machining, EDM, etc. This includes availability of high power, high beam quality, and short and ultrashort pulsed systems, as well as modern high-precision CNC stages and galvanometer mirror scanner systems that allow rapid laser–material interaction to improve process productivity, quality, and accuracy.

Laser machining of composite materials requires short beam–material interaction times (short pulse and high scanning speed) and/or short wavelengths (e.g., UV), to minimize heat effects. A multiple-pass approach is normally used for material removal. Compared with mechanical and water-jet cutting techniques, laser machining is slower, but some problems such as tool wear and water penetration can be avoided.

Negarestani and Li (2013) suggest that the challenges in laser processing are to minimize or eliminate thermal damage and maintain a high processing speed. To improve quality in laser cutting of CFRPs, techniques such as an additional coolant (water) or cryogenic aid gas, and pulsed and/or ultraviolet (UV) beam processing to depict the potential in laser machining of CFRPs were employed. It has also been described that in fiber-reinforced polymer (FRP) composites, when fibers and matrix exhibit closer thermal properties (e.g., polyester resin and aramid fiber), the composite thermal behavior can be microscopically homogenized. Thus, under laser cutting, aramid-fiber-reinforced polymer composites (AFRPs) do better. Thermal damages were discovered to be widely spread in laser cutting of CFRPs. The CFRP laser cuts were also shown to be effective in gas flow or others process parameters (e.g., output power and scanning speed). The use of oxygen as an aid gas enhanced thermal damage to the material due to oxidation. Several cuttings with a continuous wave (CW) beam fiber laser showed a significant reduction in delamination as scanning velocity increased. It is an intriguing discovery that should be addressed in future studies on laser cutting of CFRPs.

Samant and Dahotre (2009) described that the material removal during laser drilling is controlled by fusion, sublimation, evaporation, separation, plasma formation, and ablation. Yao et al. (2017) fabricated IN718/TiC nanocomposites by selective laser melting and studied the effect of heat treatment on microstructures and also tensile properties, as well as the effects of laser energy density on hardness and wear rate. Besides that, they demonstrated that the laser beam's focal diameter is 0.08 mm. The laser absorbed 0.45 for IN718 powders and 0.82 for TiC particles; as a result, they compared pure IN718 to as-built IN718/TiC, which had higher yield strength (YS) and ultimate tensile strength (UTS) but lower elongation. The crystal lattice space of the IN718/TiC nanocomposite is larger than that of pure IN718.

Moya-Muriana et al. (2020) carried out an experimental and numerical study on PA6/sepiolite nanocomposites and PLA with transmission laser welding (TLW), they also performed a tensile and shear simulation of both materials and showed that the mechanical analyses confirm the transition from ductile to brittle material at higher clay loading, and the materials break with negligible plastic deformation. XRD analysis revealed that increasing filler content resulted in higher orientation of PA6 crystals and sepiolite fibres. Optical and TLW trial results indicated it is mandatory to add a small load of commercial absorber additive (CAA) to the PA6/sepiolite nanocomposites in order to provide them with the suitable absorbent properties to carry out an efficient TLW process with polylactic acid (PLA).

The works published in literature mainly focused on laser cutting, followed by drilling and micromachining, while 3D LBM such as turning and milling are still awaiting practical use. Controlling two or more laser beams at various angles at the same time is a difficult operation during machining. Currently, the application of LBM is restricted to complex profile cutting in sheet metals; however, with the advent of sophisticated engineering materials, there is a need to improve LBM for cutting difficult-to-cut materials. The majority of the performance parameters mentioned by numerous studies include geometrical, metallurgical, and surface features such as surface roughness, taper formation, and HAZ. A significant number of the experimental works provided on LBM attempt to investigate the effects of parameter changes on quality attributes. Only a few academics have utilized scientific methods under the DOE umbrella to examine the LBM process. Unplanned experimental studies involve a plethora of unfavorable variables that influence performance variances and, as a result, provide inaccurate results.

3.2.4 Abrasive Air/Water Jet Machining

Abrasive water jet machining (AWJM) is growing in popularity, especially for cutting harder or less machinability materials, it has been utilized extensively in pattern cutting of difficult-to-cut materials such as ceramics, laminated glass, Ti sheets, MMCs, and nanocomposites. Abrasive jet machining is characterized by the use of abrasive particles within the jet to accelerate the rate of material removal well above the water jet system. AWJ machines can be utilized on a variety of soft materials, including rubbers and foam, as well as hard brittle materials like plastics, ceramics, and glass. The principal source of energy for an AWJ machine is often a multi-reciprocating pump. Because of computer-driven movements, the cutting stream can produce parts more efficiently and precisely. AWJM is a unique method of cutting materials that may be utilized in the aerospace sector. AWJM has some advantages over standard nanomaterial cutting processes, such as generating less thermal stress or heat. To improve efficiency, the procedure is smoother and requires less postprocessing. Figure 3.10 depicts a schematic of the working of a typical AWJ machine. AWJM is also used in basically all industrial sectors, including automotive, aerospace, construction, and mining, and has numerous additional potential applications (Shukla, 2013).

Figures 2.2.4 a and b compare a rough AWJM-turned Al specimen at 90 of its initial 90 impact angles to a comparable specimen at. The surface finish of the AWJM-turned Al specimen is enhanced at modest impact angles and has a superior surface finish.

Muller et al. (2017) has demonstrated with SEM images that AWJM is a potentially useful approach for cutting polymer composites. As far as extensive damage is concerned, the penetration of the abrasive in the composite layer is difficult. Delamination and substantial material loss are detected in the initial cutting stage. As far as contamination is concerned, this depends on the type of particle. AWJ technology has shown the success of cutting glass cloth, corundum, and glass bead particles without matrix contamination (Figure 3.10).

Figure 3.11a and b shows the optical micrograph of machined surfaces having two different cutting conditions. A jet cannot cut the hard fiber in the matrix, it is deflected, and the matrix is eroded. It is possible to clearly see the typical appearance of the matrix washout and protruding fiber, particularly in (a) where kinetic energy is lacking in the jet. The fiber pulled out, and the pliers were further disturbed.

Mardi et al. (2017) investigated the surface integrity of an AWJM-produced Mg-based nanocomposite. Magnesium-based MMNCs preserve high ductility while improving yield and ultimate strength. AWJM appears to be a viable machining process for 0.66 wt.% $AI_2 0_3$ nanoparticles reinforced Mg-based MMC with good surface finish and little subsurface damage. The results show that as the traverse speed rises, so does the degree of grooving by abrasive particles and irregularity in the AWJ machined surface. Similarly, with increasing traverse speed, the values of surface roughness characteristics increase.

Ramulu and Arola (1993) investigated how the unidirectional graphite/epoxy composite material was processed using WJ as well as AWJ cutting techniques. The aim of this research is a fundamental investigation into the micromechanical performance of both the fibers and matrix of a unidirectional graphite/epoxy composite

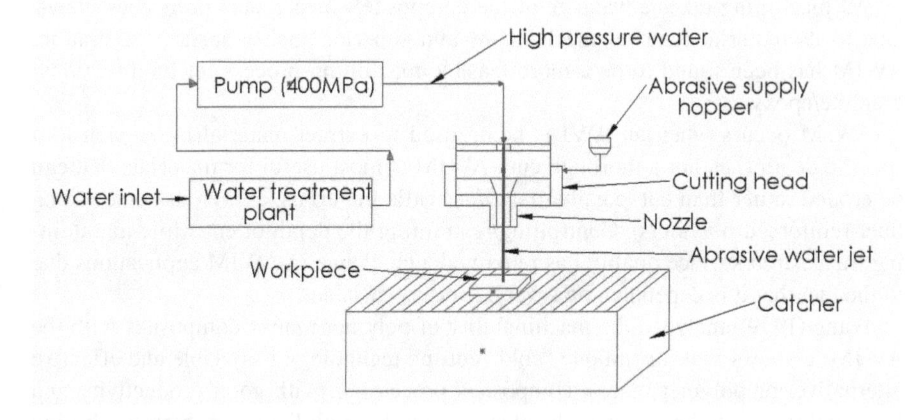

FIGURE 3.10 Schematic of a typical AWJ machine.

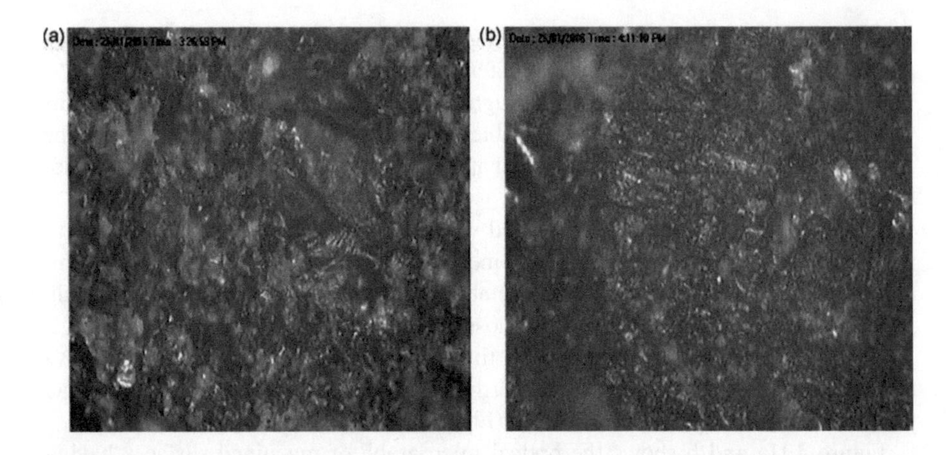

FIGURE 3.11 Optical micrograph of AWJ-machined surfaces (10x). (From Azmir and Ahsan 2008.)

material with continuous WJ and AWJ cutting circumstances. The observations will be addressed in terms of surface quality and microstructural integrity, which are both functions of fiber orientation in relation to the cutting direction.

Dimples were produced following AJM on four ceramic materials with three different abrasives under continuous machining settings. When WA abrasive was used on hard ceramics, such as Al_2O_3 and SiC, it caused slight surface roughening but no dimpling. Figure 2.2.4 depicts the development of dimples after a 10-second abrasive jet.

The feasibility of the WJ and AWJ cutting of graphite/epoxy depends on the quality of the surface it produces, both in terms of microgeometric variations and microstructural integrity, according to the discussion. In conclusion, material failure associated with microbending-induced fracture and out-of-plane shear is the primary material removal mechanism present in WJ machining of unidirectional graphite/epoxy composite (Figure 3.12).

WJ machining takes advantage of the composite's mechanical properties' flaws. Due to its material removal mechanisms and superior quality surface generation, AWJM has been found to be a more feasible machining process for unidirectional graphite/epoxy.

AWJM occurs when an AWJ is being used to extract material by erosion to a specific depth (i.e., not a thorough cut). AWJM is most useful for materials that can be eroded rather than cut (e.g., hard and/or brittle materials, as well as some strong fiber-reinforced polymers). Controlling/restraining the depth of cut while maintaining the desired surface quality has remained a challenge in AWJM applications due to the fact that it is dependent on several process variables.

Wang (1999) analyzed the machinability of polymer matrix composites with the AWJM. It shows that this unique "cold" cutting technology is a viable and effective alternative for polymer matrix composites processing, with good productivity and kerf quality, due to its distinct advantages of no thermal distortion, high machining versatility, high flexibility, and small cutting forces. It offers great potential for the

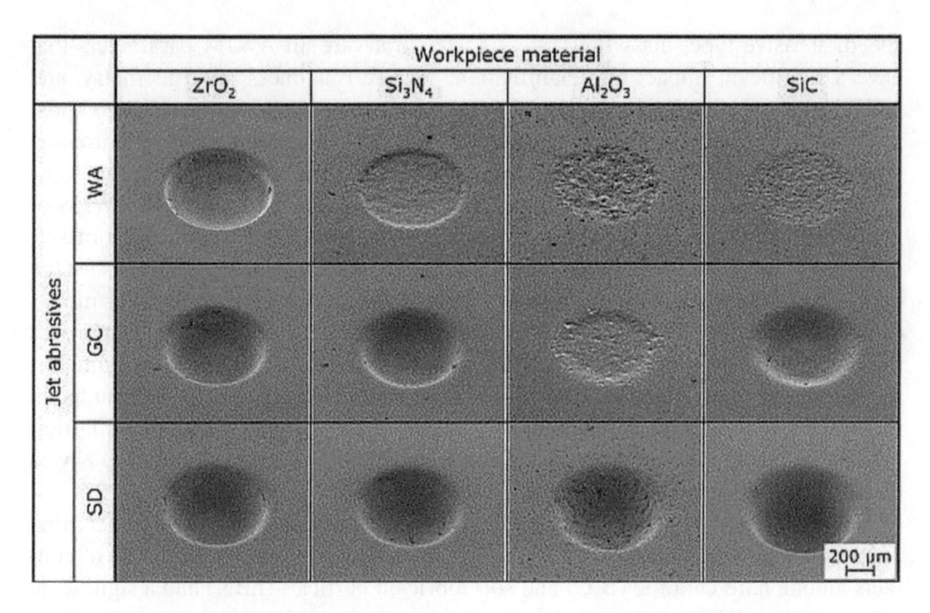

FIGURE 3.12 Appearance of the AJM face for various machining sets of abrasives and ceramic materials. (From Wakuda et al. 2002.)

processing of polymer matrix composites. It has been indicated that AWJ cutting is a feasible and effective alternative for polymer matrix composite machining with good productivity and kerf quality. Analysis and empirical modeling of kerf characteristics in terms of process parameters have provided a means for evaluating kerf geometry and compensating for kerf steepness and width in the design and processing stages.

Thirumalai Kumaran et al. (2015) studied the effect of abrasive grain size of the AWJM performance on AA(6351)-SiC-B_4C hybrid composites. Two different abrasive grain sizes (80 meshes and 120 meshes) were chosen to conduct the experiments. Cutting parameters like pressure, standoff distance, traverse speed, type of abrasive, abrasive particle size, abrasive flow rate, etc., also determined the performance, and some output responses are kerf angle, MRR, Ra, cut depth, cutting wear zone, etc. The results demonstrated that coarser abrasive particles have a positive influence on MRR; however, fine-grained abrasive particles provide a minimal kerf angle as well as excellent surface quality.

Srinivas and Babu (2012) noticed that the penetration capability of abrasive water jets in cutting aluminum-silicon carbide particulate MMCs also indicated that the share of water jet pressure in addition to traverse speed on jet penetration was greater than the abrasive flow rate.

Azmir and Ahsan (2008) examined glass-fiber-reinforced epoxy composites and discovered that abrasive hardness, operating pressure, standoff distance, and jet traversal rate are significant control parameters that impact surface roughness, and then a mathematical model was established to estimate surface roughness.

Vigneshwaran et al. (2018) investigated AWJM of fiber-reinforced composite materials and concluded that AWJM is the most viable and effective machining process for polymer composites. Stand-off distance, hydraulic pressure, transverse speed, abrasive type, mass flow rate, and feed rate are all AWJM parameters that have a significant impact on delamination, surface roughness, kerf geometry, and MRR. Water pressure and abrasive flow rate have a significant impact on composite delamination in AWJM, which cannot be avoided. When the water pressure and abrasive flow rate are increased, the kerf geometry improves.

Gnanavelbabu et al. (2018) investigated the cutting quality features of AWJM of AA6061-B_4ChBN hybrid metal matrix composites. Because it is composed of soft and hard reinforcing particles, it was found that regulating quality parameters for the hybrid metal matrix composite was more complicated and difficult than for metals or alloys. It has been seen that an increase in boron carbide process that improves the top kerf breadth when they become dislodged from the ductile aluminum surface. Reduced bottom kerf width owing to higher water energy absorption resulted in increased taper angle. Since the majority of garnet abrasive particles do not participate in the machining operation, the traverse speed has indeed been found to have a direct effect on surface roughness.

Folkes (2009a) stated that using tiny abrasive particles in AWJM machining reduces nozzle obstruction and improves machining performance. The kind of contacts among hard ceramic (B_4C) and soft lubricant particles (hBN) had a significant impact on the AWJM cutting behavior of the composite. On the ductile aluminum matrix, this causes complicated deformation as well as erosion behavior. Finally, it impacts on the cutting quality characteristics of the surface, which is the most important issue to control by altering machining settings. This work offers a thorough experimental investigation of AWJM of an AA6061-B_4C-hBN hybrid composite for evaluating machining quality parameters including kerf taper angle as well as surface roughness. The material's machining behavior was investigated using scanning electron microscopy under different process settings.

Gnanavelbabu et al. (2018) observed that increased boron carbide particles increase the width of the upper groove due to the peeling from the ductile aluminum surface, and increase the width of the lower groove due to the increased energy absorption from the water. It has been observed that the traverse speed directly affects the surface roughness as most garnet abrasive particles are not involved in the machining process. Hashish (1989) investigated the cutting and deformation wear phases in the AWJM process of ductile materials and showed that the cutting wear phase was important. Abrasive size, abrasive flow rate, water pressure, stand-off distance, and jet traverse speed are all standard AWJM process characteristics. The interactions of such process factors do have substantial impact on quality characteristics such as cutting cone angle as well as surface roughness.

Savrun and Taya (1988) focused on the machinability of metal and ceramic matrix composites using an abrasive water jet approach, and the machining performance of these high-temperature composites were addressed in terms of surface finish and microstructural integrity as a function of speed. For ceramic matrix composites, AWJ offers a promising machining technology. It is fast and usually produces smooth surfaces with little damage below the surface. AWJ also requires future improvements, such as nozzle sustainability. The machined surfaces do not exhibit microstructural changes. Results from the SiCw/A1 composite microhardness tests demonstrate that AWJ does not harden the surface. In the instance of the SiCw/A1 composite. AWJ does not cut SiC moustaches; instead, it pulls or cracks them while eroding the aluminum matrix. The impact of abrasive particles also led to localized microfusion within the matrix. There were also signs of encrustation of abrasive particles. There were no visible cracks on the machined surfaces of the SiCw/A1203 composite. Micromachining grooving was noticeable. Furthermore, plastic deformation was detected. Ramulu et al. (1993) studied the erosion properties of 30 vol.% SiC/6061-T6 Al composite and 6061-T6 Al alloy at low angles of incidence. For erosion testing, an abrasive water jet was used in the experiment, and the hydro-abrasive wear characteristics are given in terms of erosion rates, abrasive flow rate, and abrasive size. SEM has been used to characterise the eroded surface, and the results revealed that erosion in the composite occurred due to micro-cutting of the aluminium matrix material, with the Sic particulates being removed eventually by the shovelling action of the approaching jet. Finnie's erosion model was adopted and modified to predict the erosion behavior of the 3 vol% SiC/60Mt Al composite.

3.2.5 Ultrasonic Machining

Ultrasonic machining (USM) is an advanced machining method widely used for the processing of materials with higher hardness/brittleness, such as quartz, semiconductor materials, ceramics, etc., in which up to 30,000 low-energy vibration impacts are superimposed on the tool. Various parameters, such as vibration frequency, amplitude of vibration, direction of ultrasonic vibration, and cutting speed, affect the degree of cutting forces produced during the process. In comparison to thermal-based machining processes such as EDM, LBM, and so on, the machined surface created by USM is found to be free of any surface defects (heat affected region, fractures, recast coating, and so on). Aside from this, dynamic stability, reduction of cutting force, an increase in tool life, reduction of residual stress, and reduction of cutting temperature have all been reported (Patil et al. 2014).

USM has developed an ultrasonic vibration-based method for removing materials from the inside of a workpiece, using ultrasonic vibration to excite and permeate it. Figure 3.13 depicts a schematic of the key components of a typical USM process configuration. The transducer converts high-frequency electrical energy into mechanical vibration at the resonant frequency depicted in Figure 3.14. The agitated vibration is then passed through an energy-focusing horn to increase the vibration amplitude before being added to the tool tip. As a consequence, the instrument immediately above the workpiece can vibrate at a high amplitude along its longitudinal axis. A slurry of hard abrasive particles in a fluid is continuously pumped into the machining area. During the manufacturing of hard and brittle materials, numerous small fractures occur on the work surface, causing the material to be removed. Research attempts have been made to improve the machinability of ceramics nanocomposites that have a higher hardness and a lower rupture strength. Machinability can be improved by limiting the usual cutting force to minimize potential cracks, or by modifying the structure (Majeed et al. 2008).

Figure 3.14a depicts a picture of four different tool sizes (3, 1.65, 1, and 0.7) producing hole arrays on a 3.2-mm-thick SiC plate using the USM method. Throughout, all of the SiC holes were drilled. Close-up microscopic images of the entry edge of one of each size, obtained at random, are shown in Figure 3.14b–e.

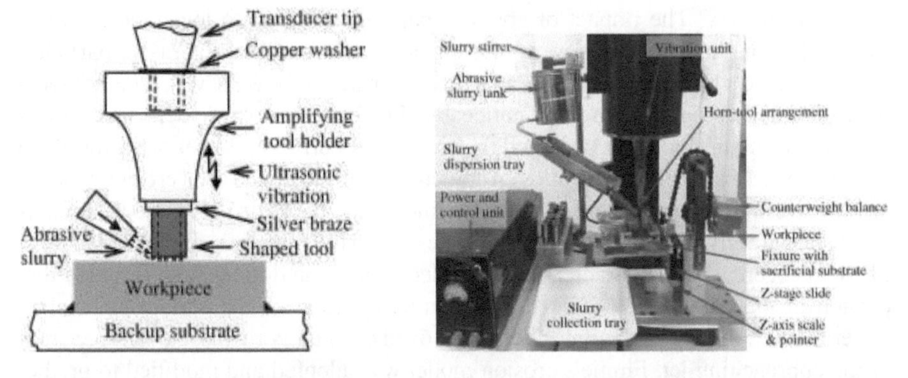

FIGURE 3.13 Schemes showing the USM approach with important features and USM configuration with integrated z-stage for the processing. (From Nath et al. 2012.)

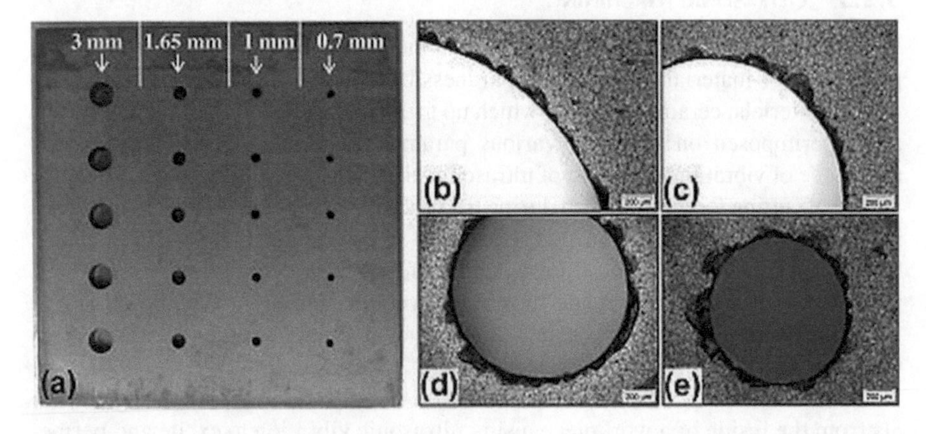

FIGURE 3.14 (a) Hole arrays produced by 3, 1.65, 1, and 0.7 mm tools on the SiC plate applying the USM process; (b–e) close-up microscopic views of the edges of one hole of each size. (From Nath et al. 2012.)

The use of USM to drill multiple holes simultaneously is shown in Figure 3.15 (Rutan 1984).

Sabyrov et al. (2019) studied the fundamentals of ultrasonic-vibration-enabled EDM measurement, its presentation limitations, ways for utilization, quantify, metallurgical highlights on a surface level, and cycle streamlining.

The machining features, like the quality of the machined surface and tool wear rate, were investigated by Kumar and Khamba (2008) in terms of the surface finish on pure titanium (ASTM Grade-I). The material removal mechanism has also been correlated with machining conditions. The optimum setting of the parameters is established by experiments using the Taguchi method. It was observed that in the ultrasonic drilling of pure titanium, the tool wear rate and the surface quality achieved are closely interlinked and, therefore, the optimization of one contributes to the optimization of the other. Figure 3.16 demonstrates the surface of an

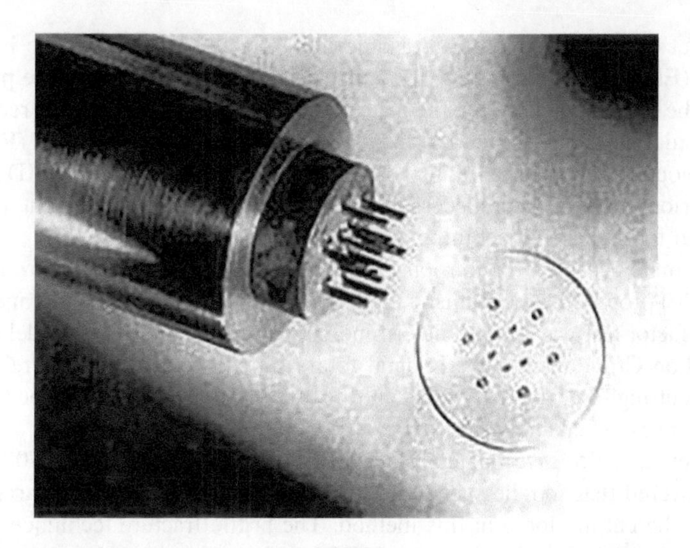

FIGURE 3.15 USM gang drilling tool for several holes. (Adapted from Singh and Khamba 2006; Rutan 1984.)

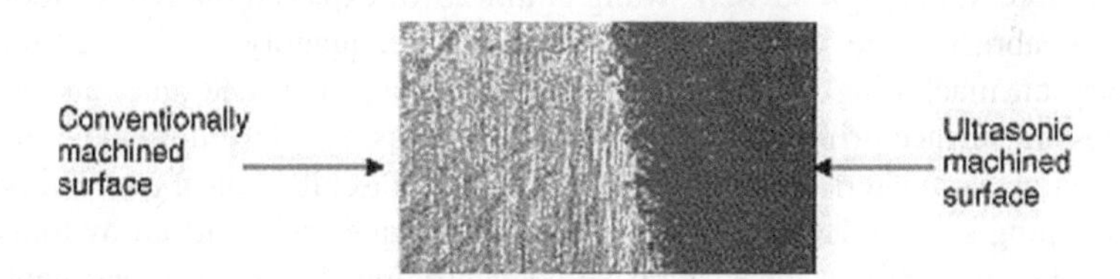

FIGURE 3.16 (i) SEM photographs reveals pulverized & brittle fracture regions during USM (From Lv et al. 2012.) (ii) Workpiece surface during USM with different coolants (a) milk, (b) water, and (c) PCD slurry (From Patwardhan 2012).

ultrasonically machined titanium sample has a non-directional surface pattern while compared to a traditionally

Cong et al. (2011, 2012) presented the findings of an experimental study on viable regions for rotary ultrasonic machining of CFRP using cold air as a coolant. Three factors – burning of the machined surface, delamination, and tool blockage – were used to assess the feasible regions in each of the input parameters (feed rate, rotational speed of the tool, ultrasonic power, and cold air pressure), and it was discovered that higher cold air pressure resulted in larger feasible regions. Dry machining was impossible when the tool rotation speed was too low, regardless of the feed rate or ultrasonic power levels.

Li et al. (2005) also drilled composite materials using rotary ultrasonic machining (RUM). Process parameters such as spindle speed, feed rate, and ultrasonic strength are investigated for their effects on responses, cutting force, MRR, chipping thickness, and chipping size. As a result, it was found that the cutting force can be reduced significantly while MRR can be increased with RUM. With the right machining parameters, this can produce high-quality holes on CMC sheets. Spindle speed and feed rate have a significant influence on hole quality.

Zhao et al. (2017) explored ultrasonic vibration-assisted electrolytic in-process dressing (ELID) mirror grinding and constructed a surface performance prediction model. The interaction between the grits and the workpiece surface was represented in ultrasonic vibration-assisted ELID grinding, and a surface roughness (Rz) model was developed. The surface quality of ultrasonic vibration-assisted ELID grinding was superior to that of its predecessor, and total variance was not evident, making it more ideal for precision machining of nanocomposite ceramics.

To estimate cutting force in rotary ultrasonic drilling of CFRP materials, Zhang et al. (2016) proposed a mechanistic model based on impression fracture mechanics. A single factor test was used to determine the parameter "K." This model was then validated on CFRP using experimental rotary ultrasonic drilling. The relationship between cutting force and its critical cutting factors, namely, spindle speed and feed rate, was investigated and addressed.

Yuan et al. (2015) investigated rotary ultrasonic drilling on CFRP-T700 material and discovered that spindle speed and feed rate are significant input variables that influence the cutting force in this method. The brittle fracture technique has been used, and a model of cutting force for CFRP-T700 was developed by using the rotary ultrasonic drilling technique. The data of cutting force collected from the model and its experimental setup were then explored, and it was discovered that the simulated and measured values agreed well. Wang et al. (2016) explored the role of tool variables, viz., abrasive size, concentration of the abrasive, quantity of slots, and tool end geometry on machining performances such as cutting power, torque, and surface roughness in surface grinding of CFRP composites using rotary ultrasonic machining. Varying parameters were observed using a data collection method linked to the machining system. Higher cutting forces and torque are produced by tools with a larger abrasive scale, a lower abrasive concentration, and convex end geometry, whereas lower surface roughness is generated by tools with greater abrasive size, a lower abrasive concentration, and convex end shape. Gopalakannan and Senthilvelan (2015) intend to evaluate the mechanical characteristics and microstructure of MMNCs made of Al 7075 matrix with 0.5 wt.% SiC and 0.5 wt.% BAC reinforcements. Further study of the aforementioned characteristics for their equivalent monolithic alloy has also been performed in order to compare them. It was discovered that the tensile strength and hardness of MMNCs have significantly improved.

Independent of material characteristics, USM has been recognized as a perfectly competent manufacturing technology for effectively processing an extensive range of advanced engineering materials, ceramics, composites, nanocomposites, etc. USM investigations have been carried out primarily to determine the process's capacity to machine different materials and the impact of process factors, and to model specific process outputs of interest. Studies to be carried out in future will face substantial restrictions in some important elements of the USM process.

There is a lack of experimental-study-based investigation with experimental design have been identified in the existing literature. Researchers discovered that nanocomposites may be machined with varying work thickness, homogeneity, anisotropy, and density. The discovery was made after examining binder phase ingredients and concentrations. It is also important to examine the correlation between the integrity of the machined hole and the removal process. In USM, the various wear processes of impregnated diamond abrasive tools must be thoroughly investigated. Potential methods of minimizing grain pull-out or dulling of abrasive diamonds may be investigated.

USM is a nontraditional hybrid manufacturing technology that is friendly to the environment. It has been used effectively in the commercial treatment of a wide variety of hard and brittle materials, together with nanocomposites, and ceramics like Zerodur, Macor, zirconia, and alumina. This technique has also been used to develop strong and ductile materials along with titanium and its alloys, stainless steel, and nanocomposites. The excellence and productivity of USM depend essentially upon the parameters of the working material, the machine's characteristics, and the machining process configurations of diamond-impregnated tools.

3.3 CHALLENGES AND FUTURE PROSPECTS OF ADVANCED MACHINING

Advanced manufacturing is important essentially for its ability to promote economic conditions through the industrialization of advanced technologies. The ability to develop new opportunities for entirely new products and services is found within new technologies. Several significant studies have been carried out to develop the machining of nanocomposites in a few cases with a large initial investment, no new design practice has been developed to take advantage of this. Nanocomposite machining can be performed by several advanced techniques, including EDM, ECM, USM, LBM, and EBM (electron beam machining). Inherent in all these processes, a final in-service geometry with the needed properties of materials is required at the cheapest price expense. Since the achievement of net form early in processing is an elusive aim. Nontraditional machining processes like EDM, ECM, WJM, AWJM, USM, and LBM provide precision finishing, yet are expensive and slow (Jahanmir et al. 1999). Consequently, machining will continue to be a problem to resolve in terms of achieving the expected strict dimensional tolerances and surface finish (Borgonovo and Apelian 2011).

These methods allow cost-effective machining of high-performance materials, which are usually difficult or impossible to machine conventionally. Environmental sustainability puts extra pressure on such emerging innovations, relating procedure development innovations. One of the significant drawbacks of EDM and miniature EDM is the low throughput/MRR. Another limitation is insecurity because of tiny flash holes (in the scope of microns), which result in long machining time, high anode wear, and surface deformities.

Several significant investigations have been made to develop the machining of nanocomposites in a few cases with a large initial investment, but no new design practice has been developed to take advantage of this. The main areas of research and development activities necessary to resolve these difficulties include the following:

- Application of advanced processing methods for machining nanocomposites
- Latest current technological improvements
- Modeling and simulation of machining processes
- Formulation of design methods capable of taking advantage of emerging technology

Micromachining development will become increasingly important in the future with the rising demand for more efficient space utilization with more competent and higher-quality goods. Research is focused on fundamental elements like material evaluation, process capability evaluation, the generation of pulsed power sources, and parametric investigations to assess the performance of the process. Certain developments, such as the manufacture of specialized machine tools and alternatives for electrolytes, are yet in their early stages and require further investigations.

It is rather challenging to drill into brittle materials. Although laser machining helps with this problem, proper hole circularity in microdrilling operations is challenging. Furthermore, because of the focusing properties of the laser beam, producing holes without tapers is quite challenging. However, it is preferable to have the drilled holes round and without taper from a production standpoint. Several researchers have explored various elements of microdrilling in different types of materials. Yet, no extensive investigation into the circularity or even taper of the laser microdrilled holes in ceramic composites has been revealed. Mostly, in industry, there is no technological assistance for laser microdrilling of these valuable materials. Consequently, understanding the circularity of microdrilled holes is critical for the successful usage of this material in the recent manufacturing sector.

Laser beams can produce cuts only 0.08 mm wide, compared with 0.5–2.5 mm using water jets. They can also cut near part edges and do not usually introduce delamination. One problem with laser cutting is that, as the thickness of the part increases, materials to be removed interfere with the laser beam and limit its ability to cut. Raising the power will result in a considerably wider HAZ and greater charring on the cut surface. The presence of a HAZ, in which fibers protrude from an overheated matrix, is cause for worry. Yet, early tests show that edge effects on thin specimens subjected to tensile stress are not a problem.

Process capability research is required to improve the process capability index for quality attributes such as surface quality. Power consumption, surface quality, depth of cut limitation value, and machining efficiency reduction are all things to look into. It is possible to use USM to improve the performance of materials with varying mechanical properties. Microhardness of the machined surface, subsurface hardness profile, and subplate damage must all be investigated for a more accurate evaluation of the surface integrity of ultrasonically machined surfaces. These concerns are critical for work materials such as titanium, which is used in typical applications that require the least amount of surface integrity loss after machining. Masked microblasting yields a minimum channel width of 10 mm. AAJM is a market leader in surface microtexture for tribological applications. AAJM has a lower MRR than liquid-based abrasive jet polishing, but it produces a smoother surface. More research into the cavitation phenomenon during air/water jet machining has the potential to improve MRR while preserving a high-quality surface. Because surface texturing is the primary goal of AAJM, the research concentrated on predicting machining profiles while ignoring surface roughness and integrity.

Fundamental elements such as material evaluation, process capability evaluation, the generation of pulsed power sources, and parametric investigations to assess process performance were the focus of research.

BIBLIOGRAPHY

Azmir, M. and Ahsan, A. (2008). Investigation on glass/epoxy composite surfaces machined by abrasive water jet machining. *Journal of Materials Processing Technology*, 198(1–3):122–128.

Biswas, R., Kuar, A., Sarkar, S., and Mitra, S. (2010). A parametric study of pulsed Nd: YAG laser micro-drilling of gamma-titanium aluminide. *Optics & Laser Technology*, 42(1):23–31.

Borgonovo, C. and Apelian, D. (2011). Manufacture of aluminum nanocomposites: a critical review. In L. Ceschini and R. Montanari (Eds.) *Materials Science Forum*, vol. 678, pp. 1–22. Trans Tech Publ, Switzerland.

Callister, W. D. and Rethwisch, D. G. (2018). *Materials Science and Engineering: an Introduction*. Wiley, New York.

Cong, W., Pei, Z. J., Deines, T. W., and Treadwell, C. (2011). Rotary ultrasonic machining of CRPF using cold air as coolant: feasible regions. *Journal of Reinforced Plastics and Composites*, 30(10):899–906.

Cong, W., Zou, X., Deines, T., Wu, N., Wang, X., and Pei, Z. (2012). Rotary ultrasonic machining of carbon fiber reinforced plastic composites: an experimental study on cutting temperature. *Journal of Reinforced Plastics and Composites*, 31(22):1516–1525.

Dubey, A. K. and Yadava, V. (2008). Laser beam machining—a review. *International Journal of Machine Tools and Manufacture*, 48(6):609–628.

Folkes, J. (2009a). Waterjet—an innovative tool for manufacturing. *Journal of Materials Processing Technology*, 209(20):6181–6189. Special Issue: 1st International Conference on Abrasive Processes.

Folkes, J. (2009b). Waterjet—an innovative tool for manufacturing. *Journal of Materials Processing Technology*, 209(20):6181–6189.

George, P., Raghunath, B., Manocha, L., and Warrier, A. M. (2004). EDM machining of carbon–carbon composite—a Taguchi approach. *Journal of Materials Processing Technology*, 145(1):66–71.

Ghany, K. A. and Newishy, M. (2005). Cutting of 1.2 mm thick austenitic stainless steel sheet using pulsed and CW Nd: YAG laser. *Journal of Materials Processing Technology*, 168(3):438–447.

Gnanavelbabu, A., Rajkumar, K., and Saravanan, P. (2018). Investigation on the cutting quality characteristics of abrasive water jet machining of aa6061-b4chbn hybrid metal matrix composites. *Materials and Manufacturing Processes*, 33(12):1313–1323.

Gommeringer, A., Schweizer, C., Kern, F., and Gadow, R. (2020). Electrical discharge machinable (y, nd) co-stabilized zirconia–niobium carbide ceramics. *Journal of the European Ceramic Society*, 40(11): 3723–3732.

Gopalakannan, S. and Senthilvelan, T. (2013). Application of response surface method on machining of al–sic nano-composites. *Measurement*, 46(8):2705–2715.

Gopalakannan, S. and Senthilvelan, T. (2015). Synthesis and characterisation of al 7075 reinforced with sic and b4c nano particles fabricated by ultrasonic cavitation method. *Journal of scientific and Industrial Research*, 74(5):281–285.

Gopalakannan, S., Senthilvelan, T., and Ranganathan, S. (2012). Modeling and optimization of EDM process parameters on machining of al 7075-b4c mmc using RSM. *Procedia Engineering*, 38:685–690.

Grigoriev, S. N., Kozochkin, M. P., Porvatov, A. N., Volosova, M. A., and Okunkova, A. A. (2019). Electrical discharge machining of ceramic nanocomposites: sublimation phenomena and adaptive control. *Heliyon*, 5(10):e02629.

Grigoriev, S. N., Volosova, M. A., Okunkova, A. A., Fedorov, S. V., Hamdy, K., Podrabinnik, P. A., Pivkin, P. M., Kozochkin, M. P., and Porvatov, A. N. (2020). Electrical discharge machining of oxide nanocomposite: nanomodification of surface and subsurface layers. *Journal of Manufacturing and Materials Processing*, 4(3):96.

Hashish, M. (1989). A model for abrasive-waterjet (AWJ) machining. *Journal of Engineering Material and Technology*, 111(2):154–162.

Hu, P., Sun, W., Fan, M., Qian, J., Jiang, J., Dan, Z., Lin, Y., Nan, C.-W., Li, M., and Shen, Y. (2018). Large energy density at high-temperature and excellent thermal stability in polyimide nanocomposite contained with small loading of batio3 nanofibers. *Applied Surface Science*, 458:743–750.

Jahan, M. P. (2019). Electro-discharge machining (EDM). In Muammer Koç and Tuğrul Özel (Ed) *Modern Manufacturing Processes*, pp. 377–409. John Wiley & Sons, Inc.

Jahanmir, S., Ramulu, M., and Koshy, P. (1999). *Machining of Ceramics and Composites*. Marcel Dekker, New York.

Jain, V., Rao, P. S., Choudhary, S., and Rajurkar, K. (1991). Experimental investigations into traveling wire electrochemical spark machining (TW-ECSM) of composites. *Journal of Engineering for Industry*, 113(1):75–84.

Jain, V. K. (2009). *Advanced Machining Processes*. Allied publishers.

Klink, A. (2016). Process signatures of EDM and ECM processes–overview from part functionality and surface modification point of view. *Procedia CIRP*, 42:240–245.

Kuar, A., Doloi, B., and Bhattacharyya, B. (2006). Modelling and analysis of pulsed Nd:YAG laser machining characteristics during micro-drilling of zirconia (zro2). *International Journal of Machine Tools and Manufacture*, 46(12):1301–1310.

Kumar, J. and Khamba, J. (2008). An experimental study on ultrasonic machining of pure titanium using designed experiments. *Journal of the Brazilian Society of Mechanical Sciences and Engineering*, 30(3):231–238.

Lee, T. and Deng, J. (2002). Mechanical surface treatments of electro-discharge machined (EDMed) ceramic composite for improved strength and reliability. *Journal of the European Ceramic Society*, 22(4):545–550.

Li, Z., Jiao, Y., Deines, T., Pei, Z., and Treadwell, C. (2005). Rotary ultrasonic machining of ceramic matrix composites: feasibility study and designed experiments. *International Journal of Machine Tools and Manufacture*, 45(12–13):1402–1411.

Liu, C.-C. and Huang, J.-L. (2003). Effect of the electrical discharge machining on strength and reliability of tin/si3n4 composites. *Ceramics International*, 29(6):679–687.

Liu, J. and Yue, T. (2012). Electrochemical discharge machining of particulate reinforced metal matrix composites. In H. Hocheng (Ed.) *Machining Technology for Composite Materials*, pp. 242–265. Woodhead Publishing Limited, Cambridge, UK.

Lok, Y. and Lee, T. (1997). Processing of advanced ceramics using the wire-cut EDM process. *Journal of Materials Processing Technology*, 63(1):839–843.

Lv, D., Wang, H., Tang, Y., Huang, Y., Zhang, H., and Ren, W. (2012). Surface observations and material removal mechanisms in rotary ultrasonic machining of brittle material. *Proceedings of the Institution of Mechanical Engineers, Part B: Journal of Engineering Manufacture*, 226(9):1479–1488.

Mahanta, S., Chandrasekaran, M., and Samanta, S. (2019). RSM modeling and Taguchi analysis of EDM of B4C and flyash reinforced hybrid nanocomposites. In *Composite Materials and Material Engineering III*, vol. 801 of Key Engineering Materials, pp. 227–232. Trans Tech Publications Ltd, Switzerland.

Majeed, M. A., Vijayaraghavan, L., Malhotra, S., and Krishnamurthy, R. (2008). Ultrasonic machining of Al2O3/LaPO4 composites. *International Journal of Machine Tools and Manufacture*, 48(1):40–46.

Mardi, K. B., Dixit, A., Mallick, A., Pramanik, A., Ballokova, B., Hvizdos, P., Foldyna, J., Scucka, J., Hlavacek, P., and Zelenak, M. (2017). Surface integrity of mg-based nanocomposite produced by abrasive water jet machining (awjm). *Materials and Manufacturing Processes*, 32(15):1707–1714.

Mathiazhagan, S. and Anup, S. (2018). Atomistic simulations of length-scale effect of bio-inspired brittle-matrix nanocomposite models. *Journal of Engineering Mechanics*, 144(11):04018104.

Mausam, K., Sharma, K., Bharadwaj, G., and Singh, R. P. (2019). Multi-objective optimization design of die-sinking electric discharge machine (EDM) machining parameter for CNT-reinforced carbon fibre nanocomposite using grey relational analysis. *Journal of the Brazilian Society of Mechanical Sciences and Engineering*, 41(8):1–8.

Melk, L., Antti, M.-L., and Anglada, M. (2016). Material removal mechanisms by EDM of zirconia reinforced mwcnt nanocomposites. *Ceramics International*, 42(5):5792–5801.

Mohan, B., Rajadurai, A., and Satyanarayana, K. (2002). Effect of sic and rotation of electrode on electric discharge machining of al–sic composite. *Journal of Materials Processing Technology*, 124(3):297–304.

Moya-Muriana, J. A., Yebra-Rodrguez, A., La Rubia, M. D., and Navas-Martos, F. J. (2020). Experimental and numerical study of the laser transmission welding between pa6/sepiolite nanocomposites and pla. *Engineering Fracture Mechanics*, 238: 107277.

Muller, M., D'armato, R., and Rudawska, A. (2017). Machining of polymeric composites by means of abrasive water-jet technology. In *16th International Scientific Conference Engineering for Rural Development. Jelgava, Latvia University of Agriculture*, pp. 121–127.

Nath, C., Lim, G., and Zheng, H. (2012). Influence of the material removal mechanisms on hole integrity in ultrasonic machining of structural ceramics. *Ultrasonics*, 52(5):605–613.

Negarestani, R. and Li, L. (2012). Laser machining of fibre-reinforced polymeric composite materials. In H. Hocheng (Ed.) *Machining Technology for Composite Materials*, pp. 288–308. Woodhead Publishing Limited, Cambridge, UK.

Negarestani, R. and Li, L. (2013). Fibre laser cutting of carbon fibre–reinforced polymeric composites. *Proceedings of the Institution of Mechanical Engineers, Part B: Journal of Engineering Manufacture*, 227(12):1755–1766.

Negarestani, R., Li, L., Sezer, H., Whitehead, D., and Methven, J. (2010). Nanosecond pulsed DPSS Nd: YAG laser cutting of CFRP composites with mixed reactive and inert gases. *The International Journal of Advanced Manufacturing Technology*, 49(5–8):553–566.

Patel, K., Pandey, P. M., and Rao, P. V. (2009). Surface integrity and material removal mechanisms associated with the EDM of al2o3 ceramic composite. *International Journal of Refractory Metals and Hard Materials*, 27(5):892–899.

Patel, K., Pandey, P. M., and Rao, P. V. (2010). Optimisation of process parameters for multi-performance characteristics in EDM of al 2 o 3 ceramic composite. *The International Journal of Advanced Manufacturing Technology*, 47(9–12):1137–1147.

Patel, K., Pandey, P. M., and Rao, P. V. (2011). Study on machinability of al2o3 ceramic composite in EDM using response surface methodology. *Journal of Engineering Materials and Technology*, 133(2):021004.

Patil, S., Joshi, S., Tewari, A., and Joshi, S. S. (2014). Modelling and simulation of effect of ultrasonic vibrations on machining of ti6al4v. *Ultrasonics*, 54(2):694–705.

Patwardhan, A. V. (2012). *Experimental investigation of hard and brittle materials machining using micro rotary ultrasonic machining*. PhD thesis, University of Nebraska.

Ponappa, K., Aravindan, S., Rao, P., Ramkumar, J., and Gupta, M. (2010). The effect of process parameters on machining of magnesium nano alumina composites through EDM. *The International Journal of Advanced Manufacturing Technology*, 46(9–12):1035–1042.

Pradhan, M. (2013). Estimating the effect of process parameters on MRR, TWR and radial overcut of EDMed AISI D2 tool steel by RSM and GRA coupled with PCA. *International Journal of Advanced Manufacturing Technology*, 68(1–4):591–605.

Pradhan, M. K. (2010). *Experimental investigation and modelling of surface integrity, accuracy and productivity aspects in EDM of AISI D2 steel.* PhD thesis, National Institute of Technology, Rourkela.

Pradhan, M. K. (2020). Tribological behaviour, machinability, and optimization of EDM of aa-2014 hybrid composite reinforced with sic and glass particulates. In Prasanta Sahoo (Ed.) *Handbook of Research on Developments and Trends in Industrial and Materials Engineering*, vol. 1, pp. 228–269. IGI Global, USA.

Pradhan, M. K. and Biswas, C. K. (2010). Neuro-fuzzy and neural network-based prediction of various responses in electrical discharge machining of AISI D2 steel. *The International Journal of Advanced Manufacturing Technology*, 50(5–8):591–610.

Pradhan, M. K. and Dehari, A. (2019). Optimization of process parameters for electrical discharge machining of al7075, b4c and tic hybrid composite using ELECTRE method. In *Optimization using Evolutionary Algorithms and Metaheuristics: Applications in Engineering*, vol. 1, pp. 57–80. CRC Press Taylor and Francis Groups, London.

Pradhan, M. K. and Singh, B. (2019). Machinability and multi-response optimization of EDM of al7075/sic/ws2 hybrid composite using the PROMETHEE method. In *Optimization for Engineering Problems*, vol. 1, pp. 39–75. John Wiley & Sons, Inc, London, UK.

Pradhan, M. K. and Tiwari, S. (2017). Investigation of mechanical properties and electrical discharge machining of lm25-rha metal matrix composite. *International Journal of Machining and Machinability of Materials*, 19(5):457–482.

Prakash, J. and Gopalakannan, S. (2020). Teaching—learning-based optimization coupled with response surface methodology for micro electrochemical machining of aluminium nanocomposite. *Silicon*, 13:1–24.

Quarto, M., Bissacco, G., and D'Urso, G. (2019). Machinability and energy efficiency in micro-EDM milling of zirconium boride reinforced with silicon carbide fibers. *Materials*, 12(23):3920.

Rajkumar, K., Poovazhgan, L., Saravanamuthukumar, P., Javed Syed Ibrahim, S., and Santosh, S. (2015). Abrasive assisted electrochemical machining of al-b4c nanocomposite. In *Applied Mechanics and Materials,* vol.787, pp. 523–527. Trans Tech Publ, Switzerland.

Ramulu, M. and Arola, D. (1993). Water jet and abrasive water jet cutting of unidirectional graphite/epoxy composite. *Composites*, 24(4):299–308.

Ramulu, M., Raju, S., Inoue, H., and Zeng, J. (1993). Hydro-abrasive erosion characteristics of 30 vol.% SiCp/6061-T6 Al composite at shallow impact angles. *Wear*, 166(1):55–63.

Rao, S. and Padmanabhan, G. (2014). Optimization of machining parameters in ECM of al/b4c composites using Taguchi method. *International Journal of Applied Science and Engineering*, 12(2):87–97.

Rao, S. R. and Padmanabhan, G. (2013). Optimization of machining parameters in ECM of al/b4c composites. *Journal for Manufacturing Science and Production*, 13(3):145–153.

Rathod, V. T., Kumar, J. S., and Jain, A. (2017). Polymer and ceramic nanocomposites for aerospace applications. *Applied Nanoscience*, 7(8):519–548.

Roy, N., Kuar, A., Mitra, S., and Acherjee, B. (2015). Nd: Yag laser microdrilling of SiC-30BN nanocomposite: experimental study and process optimization. In *Lasers Based Manufacturing*, pp. 317–341. Springer, New Delhi.

Rutan, H. (1984). Ultrasonic machining (impact grinding). Technical report, Lawrence Livermore National Lab., CA (USA).

Sabyrov, N., Jahan, M., Bilal, A., and Perveen, A. (2019). Ultrasonic vibration assisted electro-discharge machining (EDM)—an overview. *Materials*, 12(3):522.

Samant, A. N. and Dahotre, N. B. (2009). Laser machining of structural ceramics—a review. *Journal of the European ceramic society*, 29(6):969–993.

Savrun, E. and Taya, M. (1988). Surface characterization of SiC whisker/2124 aluminium and Al2O3 composites machined by abrasive water jet. *Journal of materials science*, 23(4):1453–1458.

Saxena, K. K., Qian, J., and Reynaerts, D. (2018). A review on process capabilities of electrochemical micromachining and its hybrid variants. *International Journal of Machine Tools and Manufacture*, 127:28–56.

Senthilkumar, C., Ganesan, G., and Karthikeyan, R. (2009). Study of electrochemical machining characteristics of al/sic p composites. *The International Journal of Advanced Manufacturing Technology*, 43(3–4):256–263.

Shukla, M. (2013). Abrasive water jet milling. In J. Paulo Davim (Ed) *Non-traditional Machining Processes*, pp. 177–203. Springer, London.

Singh, N. B. and Agarwal, S. (2016). Nanocomposites: an overview. *Emerging Materials Research*, 5(1):5–43.

Singh, P. N., Raghukandan, K., Rathinasabapathi, M., and Pai, B. (2004). Electric discharge machining of al–10% SiCp as-cast metal matrix composites. *Journal of Materials Processing Technology*, 155:1653–1657.

Singh, R. and Khamba, J. (2006). Ultrasonic machining of titanium and its alloys: a review. Journal of Materials Processing Technology, 173(2):125–135.

Snoeys, R., Staelens, F., and Dekeyser, W. (1986). Current trends in nonconventional material removal processes. *CIRP Annals – Manufacturing Technology*, 35(2):467–480.

Soni, S. and Thomas, B. (2017). A comparative study of electrochemical machining process parameters by using GA and Taguchi method. In Sergey Prokoshkin, Natalia Resnina, Sergey Dubinskiy, Yulia Zhukova, Vadim Sheremetyev, Victor Komarov, Kristina Polyakova (Eds.), *IOP Conference Series: Materials Science and Engineering*, vol. 263, p. 062038. IOP Publishing, Bristol, UK.

Srinivas, S. and Babu, N. R. (2012). Penetration ability of abrasive waterjets in cutting of aluminum-silicon carbide particulate metal matrix composites. *Machining Science and Technology*, 16(3):337–354.

Tavangarian, F., Fahami, A., Li, G., Kazemi, M., and Forghani, A. (2018). Structural characterization and strengthening mechanism of forsterite nanostructured scaffolds synthesized by multistep sintering method. *Journal of Materials Science & Technology*, 34(12):2263–2270.

Thirumalai Kumaran, S., Uthayakumar, M., Mathiyazhagan, P., Krishna Kumar, K., and Muthu Kumar, P. (2015). Effect of abrasive grain size of the AWJM performance on aa (6351)-SiC-B4C hybrid composite. In K. Palanikumar (Ed.), *Applied Mechanics and Materials*, vol.766, pp. 324–329. Trans Tech Publ, Switzerland.

Tiwari, S. and Pradhan, M. (2017). Optimisation of machining parameters in electrical discharge machining of LM-25-RHA composites. In Raja Das and Mohan K. Pradhan (Eds.), *Handbook of Research on Manufacturing Process Modeling and Optimization Strategies*, pp. 1–18. IGI Global, Hershey, PA.

Vigneshwaran, S., Uthayakumar, M., and Arumugaprabu, V. (2018). Abrasive water jet machining of fiber-reinforced composite materials. *Journal of Reinforced Plastics and Composites*, 37(4):230–237.

Volosova, M. A., Okunkova, A. A., Povolotskiy, D. E., and Podrabinnik, P. A. (2015). Study of electrical discharge machining for the parts of nuclear industry usage. *Mechanics & Industry*, 16(7):706.

Wakuda, M., Yamauchi, Y., and Kanzaki, S. (2002). Effect of workpiece properties on machinability in abrasive jet machining of ceramic materials. *Precision Engineering*, 26(2):193–198.

Wang, H., Ning, F., Hu, Y., Fernando, P., Pei, Z. J., and Cong, W. (2016). Surface grinding of carbon fiber–reinforced plastic composites using rotary ultrasonic machining: effects of tool variables. *Advances in Mechanical Engineering*, 8(9):1687814016670284.

Wang, J. (1999). Abrasive waterjet machining of polymer matrix composites–cutting performance, erosive process and predictive models. *The International Journal of Advanced Manufacturing Technology*, 15(10):757–768.

Wu, N., Li, X., Li, J.-G., Zhu, Q., and Sun, X. (2018). Fabrication of gD2o3mgo nanocomposite optical ceramics with varied crystallographic modifications of gD2o3 constituent. *Journal of the American Ceramic Society*, 101(11):4887–4891.

Xu, Z. and Wang, Y. (2019). Electrochemical machining of complex components of aero-engines: developments, trends, and technological advances. *Chinese Journal of Aeronautics*, 34(2):28–53.

Yadav, P., Yadava, V., and Narayan, A. (2018). Experimental investigation of kerf characteristics through wire electrochemical spark cutting of alumina epoxy nanocomposite. *Journal of Mechanical Science and Technology*, 32(1):345–350.

Yadav, P., Yadava, V., and Narayan, A. (2020). Experimental investigation for performance study of wire electrochemical spark cutting of silica epoxy nanocomposites. *Silicon*, 12:1023–1033.

Yao, X., Moon, S. K., Lee, B. Y., and Bi, G. (2017). Effects of heat treatment on microstructures and tensile properties of in718/tic nanocomposite fabricated by selective laser melting. *International Journal of Precision Engineering and Manufacturing*, 18(12):1693–1701.

Yilbas, B., Davies, R., and Yilbas, Z. (1992). Study into penetration speed during CO2 laser cutting of stainless steel. *Optics and Lasers in Engineering*, 17(2):69–82.

Yuan, S., Zhang, C., Amin, M., Fan, H., and Liu, M. (2015). Development of a cutting force prediction model based on brittle fracture for carbon fiber reinforced polymers for rotary ultrasonic drilling. *The International Journal of Advanced Manufacturing Technology*, 81(5–8):1223–1231.

Yun, J. and Wang, D. (1998). Electrical discharge machining of aluminium oxide matric composites containing titanium carbide as a conductive second phase. *Book Institute of Materials*, 680:1773–1782.

Zeller, F., Muller, C., Miranzo, P., and Belmonte, M. (2017). Exceptional micromachining performance of silicon carbide ceramics by adding graphene nanoplatelets. *Journal of the European Ceramic Society*, 37(12):3813–3821.

Zhang, C., Yuan, S., Amin, M., Fan, H., and Liu, Q. (2016). Development of a cutting force prediction model based on brittle fracture for c/Sic in rotary ultrasonic facing milling. *The International Journal of Advanced Manufacturing Technology*, 85(14):573–583.

Zhao, B., Chen, F., Jia, X.-f., Zhao, C.-y., and Wang, X.-b. (2017). Surface quality prediction model of nano-composite ceramics in ultrasonic vibration-assisted ELID mirror grinding. *Journal of Mechanical Science and Technology*, 31(4): 1877–1884.

4 Intelligent Machining of Nanocomposites

CONTENTS

4.1 INTRODUCTION

The aim of manufacturing industries is to produce high-quality products with minimum time and cost, minimum effort, minimum damage to the machine-tool-fixture and workpiece, minimum power/energy consumption, less tool wear/higher tool life, higher material removal rate/productivity, and less negative environmental effects through machining, i.e., effectively, efficiently, and economically. Machining is considered a very complex metal removal process that comes with various nonlinear and multivariate problems. Thus, an automatic control system is essential to monitor the machining process with the help of sensors that transmit feedback to the system to

DOI: 10.1201/9781003107743-4

make proper decisions and ensure optimal parametric/operating conditions. Thus, an automated system should have a model that ensures appropriate decision-making on machining process, such as finite element method (FEM). Modeling is usually performed by computational tools/soft computing tools through artificial neural networks (ANN) and fuzzy logic. Like modeling, process optimization can be performed through conventional optimization algorithm or heuristic soft algorithm called genetic algorithm (GA), simulated annealing (SA), ant colony optimization, particle swarm optimization (PSO), and process control through fuzzy logic or a combination of hard and soft techniques, called hybrid techniques.

Adaptive control plays vital role in intelligent machines to adapt to dynamic changes in the system that occur in machining processes due to variation in the hardness of workpiece, process parameters, tool wear, vibration, etc., and enhance the machining process's intelligence, i.e., capacity to acquire and apply knowledge. An intelligent machining system is also equipped to relate or monitor the faults and their effects/diagnostics through various signals that correlate with the machining process. The control and monitoring algorithms should be based on simultaneous measurement and processing of various signals [1]. In an intelligent machining center, an adaptive controller adjusts the feed rate, spindle speed, and tool path according to cutting condition variations. There are different types of adaptive control systems, such as adaptive control optimization, adaptive control constraint, geometric adaptive control, and vibration adaptive control. The advancement in and potential of intelligent machining systems have been possible due to the integration of analytical models with optimization techniques and the development of sensor-equipped machinery. The role of an analytical model is to simulate the machining process with a mathematical approach to minimize errors and use sensors to collect information in real time, which helps to yield optimal solutions without much downtime [2]. Thus, intelligent machining systems include intelligent planning, intelligent scheduling, intelligent process control, and optimization of machining parameters. This also include tool condition monitoring and chatter reduction due to integration of sensors with the cutting tool, called smart cutting tool, typically used for the overall improvement of productivity and process monitoring.

Different computational approaches and optimization methods for the development of intelligent machining system are discussed, including soft computing techniques such as neural networks in modeling, fuzzy-set-based modeling, hybrid neuro-fuzzy modeling, finite element method (FEM), genetic algorithm (GA), etc. Metal machining involves two major stages: modeling and optimization. Modeling is basically done to establish a correlation between input and output parameters whereas optimization defines the optimal parameters to achieve the desired output [3]. Basically neural networks are classified based on their architecture and training, and some widely used architectures are feed-forward neural networks, feedback neural networks, and self-organizing neural networks. Out of these, feed-forward neural networks are popularly used and are of two types: multilayer perception (MLP) neural networks and radial base function (RBF) neural networks [1]. In the feed-forward neural network architecture, each layer consists of a number of neurons and has an interconnection to the next layer. The first and second layers are called input layer and hidden layer. More than two hidden layers exist in MLP whereas one hidden layer exists in RBF. The last layer is called output layer, whose neurons constitute the response to an input pattern of the neural network. ANN needs to be trained to yield accurate response to a given input vector. It is an iterative process that adjusts the weights and biases (parameters) until it produces the desired results from the inputs.

By taking input data, a fuzzy-set-based prediction system carries out fuzzification, i.e., it undergoes translation to linguistic terms, which is then sent to an inference engine. The output of the inference system in linguistic form will undergo a defuzzification process and gets converted to numerical data [3]. In GA, a point in search space is denoted by binary or decimal numbers called chromosomes, where a fitness value is assigned to it. A set of chromosomes is known as a population, which is operated by operations such as reproduction, crossover and mutation, and it constitutes one generation. The process continues until accuracy is attained by the system after many generations [3]. The finite element method (FEM) is a widely used numerical method for the simulation of machining processes, such as temperature distribution, residual stress and strain, chip formation, and cutting force prediction.

4.2 COMPUTATIONAL TECHNIQUES

Various researchers in the recent past investigated the application of computational approaches and optimization methods for the development of intelligent machining systems, including soft computing techniques such as neural networks in modeling, fuzzy-set-based modeling, hybrid neuro-fuzzy modeling, FEM, and GA. Findings are presented for conventional and nonconventional machining processes with due emphasis on composites.

4.2.1 Artificial Neural Network

Zinati and Razfer [4] investigated end milling of PA6/nano-$CaCO_3$ composites and studied the effect of milling parameters, such as spindle speed and feed per tooth and nano-$CaCO_3$ content, on surface roughness (Ra) and total cutting force (Fw) by developing a Harmony-search-based neural network (HSNN) predictive model and analysis of variance (ANOVA). The HS algorithm has been used for the training of the feed-forward neural network and adjusts the weights and biases of MLP. From experimental observations, it can be seen that surface roughness increases with increase in feed rate, whereas spindle speed and nano-$CaCO_3$ content do not have much effect. Due to the lubricating properties of nano-$CaCO_3$ particles, cutting forces are considerably reduced for nanocomposites compared to pure PA6 (polyamide). Cutting force increases with feed rate but decreases with spindle speed. For the development of a prediction model for the responses (cutting force and surface roughness), two neural networks with two hidden layers with three inputs and one output were used. It was observed that the HSNN model is effective and reliable in modeling responses for end milling of PA6/nano-$CaCO_3$ composites due to low error in the prediction model and that it can be utilized for the optimization of the process. Teti et al. [5] analyzed multisensor process monitoring during drilling of carbon-fiber-reinforced polymer (CFRP) and CFRP composite material stacks laminate for aerospace assembly applications based on thrust force and torque signal detection to assess tool wear. ANN of the machine learning system was used to take smart decisions for appropriate changes in the cutting tool, which is essential for CFRP drilling process automation and thus can be utilized for online tool wear monitoring. Asadi et al. [6] developed an ANN model of responses such as grain size and hardness of AZ91/SiC nanocomposite during friction stir processing (FSP) with respect to FSP input parameters like traverse speed, rotational speed, and region type. Linear regression analysis was carried out to check the adequacy of the ANN model by computing correlation coefficients. The predicted results were in good agreement with experimental results, and a parametric effect on model responses was carried out by sensitivity analysis. Ray et al. [7] studied the tribomechanical performance of glass-epoxy hybrid composites filled with marble powder and optimized the process parameters for minimum erosion wear using the Taguchi approach. The ANN predictive model is found to be effective for the correlation of erosion wear performance as the experimental and predicted values are very close to each other. Bagci and Isik [8] investigated the orthogonal turning of unidirectional glass-fiber-reinforced plastics (GFRP) using full factorial design. Surface roughness models were developed using ANN and response surface (RS) with good agreement between predicted and experimental results. Maximum test errors have been found for the ANN model compared to the RS model; it also involves more computational time. Abdul Budan [9] investigated the machinability of FRP composites and developed an ANN model for cutting forces, which was obtained from finite element analysis. Furthermore, ANN results were compared with the experimental results and were found to show good accuracy. Basheer et al. [10] developed an ANN-based model for the prediction of surface roughness during precision machining of Al/SiCp composites and found a good agreement with the experimental data set, with a correlation coefficient of 97.7% and a mean error of 10.4%. It uses a feed-forward network consisting of Bayesian regularization with the Levenberg–Marquardt modification to train the neural network. A comparison of the experimental and predicted results is shown in Figure 4.1 [10].

Muthukrishnan and Davim [11] optimized the machining parameters of Al/SiC-MMC with ANOVA and ANN techniques using polycrystalline diamond (PCD) inserts during machining. ANOVA revealed that the feed rate has the highest physical as well as statistical influence on the surface roughness (51%) right after the depth of cut (30%) and the cutting speed (12%). The results of this neural network model show a close match between the model output and the directly measured surface roughness. This method seems to have prediction potentials for nonexperimental pattern additionally. The ANN methodology consumes lesser time giving higher accuracy. Multilayered perception and back propagation algorithm were used to train the network, and the validation of the experimental and ANN-predicted results is shown in Figures 4.2 [11] and 4.3 [11], respectively.

The percentage of error between them is found to be minimum and lies between 0.39–4.78. Hence, optimization using ANN is the most effective method compared with ANOVA. Panda et al. [12] investigated machinability during the machining of bearing steel for multi-responses such as flank wear, surface roughness, and chip morphology. Gray relational analysis (GRA) has been used for the optimization of multi-responses which greatly improved the methodology. Prediction models have been developed through quadratic regression model and ANN model. ANN

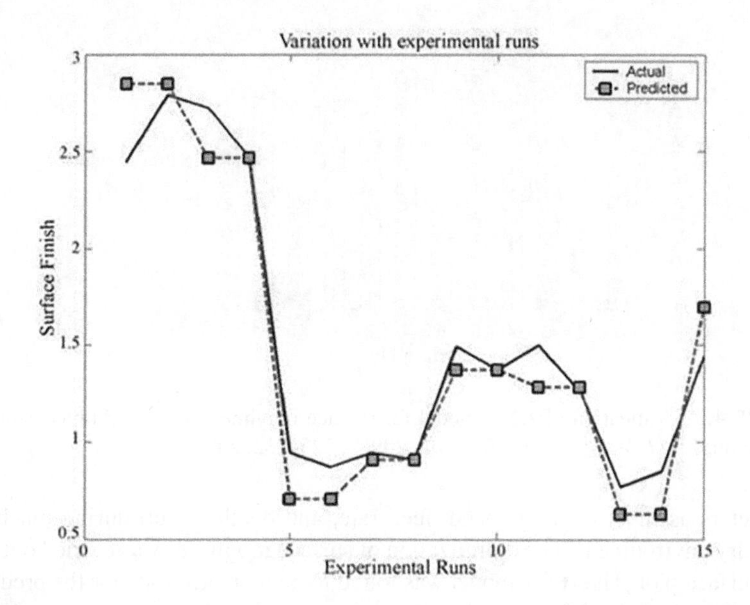

FIGURE 4.1 Comparison of predicted surface roughness with the unexposed experimental values of Al/SiC/20p/220. (From Basheer, A.C. et al., *J. Mat. Process. Technol.*, 197(1–3), 439–444, 2008.)

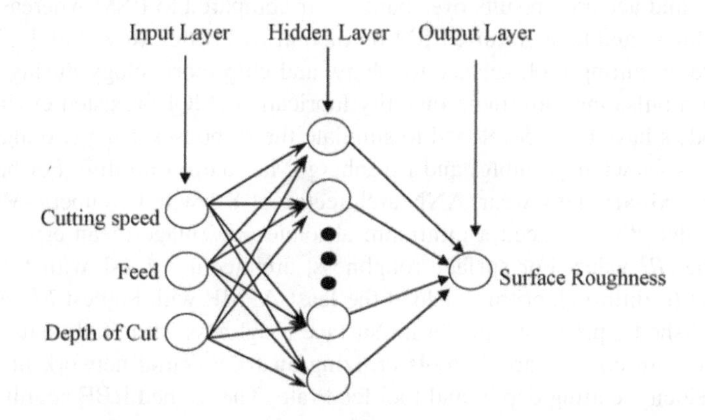

FIGURE 4.2 Configurations of neural networks. (From Muthukrishnan, N. and Davim, J.P., *J. Mat. Process. Technol.*, 209(1), 225–232, 2009.)

by multilayered feed-forward network was found to be accurate in the prediction of responses due to minimum error percentage compared to the regression model. Kumar et al. [13] developed a flank wear and chip reduction coefficient model utilizing an ANN model through feed-forward back propagation network using the Levenberg–Marquardt modification during spray-cooling-assisted machining and found it to be adequate. Sahoo et al. [14] studied the two predicted models – response surface methodology (RSM) and ANN model – for surface roughness with cutting

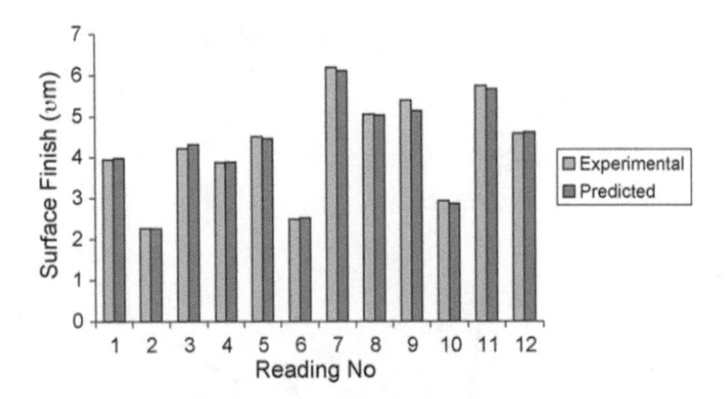

FIGURE 4.3 Validation of ANN model for surface roughness. (From Muthukrishnan, N. and Davim, J.P., *J. Mat. Process. Technol.*, 209(1), 225–232, 2009.)

parameters as input (cutting speed, feed rate, and depth of cut) during machining under dry environment. The optimization of surface roughness was carried out using a 3D surface plot. The ANN model was found to be more accurate for the prediction of response than the RSM model as the maximum error between the experimental and ANN model lies between −1.27%–0.02%. Kumar et al. [15] studied the machinability aspects considering surface roughness, flank wear, and chip–tool interface temperature with the development RSM and ANN. ANN was observed to give better prediction and accurate results over flank wear compared to RSM whereas for surface roughness and temperature, RSM prediction was better. Roy et al. [16] studied the failure of cutting tool, surface topology, and chip morphology during machining under a pulsating, minimum quantity lubrication (MQL)-assisted environment. ANN models have been developed to simulate the responses using two algorithms: BFGC quasi-Newton (trainbfg) and Livenberg–Marquardt (trainlm). For both principal and auxiliary flank wear, ANN architecture 4-8-1 with Livenberg–Marquardt (trainlm) algorithm induced a minimum absolute percentage mean error (APME) with higher R^2 value. For surface roughness, architecture 4-4-1 with Livenberg–Marquardt (trainlm) algorithm yielded the least APME with highest R^2 values. Lu [17] established a predictive model for surface roughness with 304L stainless steel workpiece with coated carbide tools utilizing an RBF neural network in terms of cutting velocity, cutting depth, and tool feed rate. The trained RBF neural network with the adaptive-adjusting variable has been observed to be the perfect network for the prediction of surface profiles. With good accuracy to the real profile, the amplitude of vibration, surface profile shape, and trend of surface profile machined by turning process. Davim et al. [18] generated a predictive model for surface roughness parameters (Ra and Rt) using ANN considering machining parameters such as cutting speed, tool feed rate, and depth of cut during turning of free machining steel. The tests were planned according to the L_{27} Taguchi OA (Orthogonal Array) to generate the knowledge base for ANN training. The study revealed that the cutting speed and tool feed rate have a significant effect on the reduction of surface roughness, whereas cutting depth has a negligible effect. Escamilla et al. [19] developed a surface roughness model using neural network and maximum sensitivity network to predict the surface roughness in turning grade 5 titanium alloy with physical vapor deposition (PVD) (TiAIN)-coated carbide inserts considering machining parameters such as cutting speed, tool feed rate, and depth of cut. It is evident that both the neural networks (back propagation and maximum sensibility) successfully analyze the variables to model the machining process. Karayel [20] employed ANN for the development of a prediction model for surface roughness parameters (Ra, Rz and Rmax) during machining of St 50.2 steel using cutting speed, depth of cut, and tool feed rate as input variables in a computer numerical control lathe. A feed-forward multilayered neural network was developed and trained using a scaled conjugate gradient algorithm (SCGA). The values of surface roughness correspond to the machining parameters, and the machining parameters for a certain roughness value can be estimated before metal cutting operation through ANN and control algorithm. Asiltürk and Çunkas [21] established surface roughness prediction models during machining of AISI 1040 steel using the full factorial design approach. Multiple regression approaches and ANN are utilized to develop surface roughness model for various tool feed rates, cutting speed, and depth of cut and were observed to be capable of prediction of response. Multilayered perception structured feed-forward ANN was used to develop the surface roughness model shown in Figure 4.4 [21].

The comparison results at the training and testing stage are shown in Figures 4.5 [21] and 4.6 [21], respectively. It has been reported that the ANN model predicted surface roughness accurately compared with multiple regression model, as statistical value, mean squared error (MSE), mean absolute error, and correlation coefficient were in the acceptable range.

Risbood et al. [22] generated an ANN model for surface roughness and can be predicted with good accuracy considering acceleration of radial vibration during turning steel. Surface roughness increases with increase in feed rate and then decreases with further rise of feed rate during turning with TiN-coated carbide cutting tool. However, this trend is different with high-speed steel cutting tool. For dimensional deviation prediction, radial component of cutting force and acceleration of radial vibration were considered. Ozel and Karpat [23] employed regression and ANN model for prediction of average surface roughness and tool wear pattern through

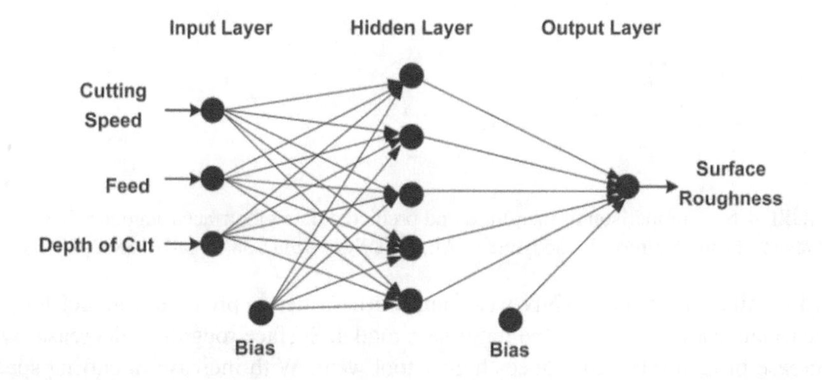

FIGURE 4.4 ANN structure. (From Asiltürk, I. and Çunkaş, M., *Exp. Syst. Appl.*, 38(5), 5826–5832, 2011.)

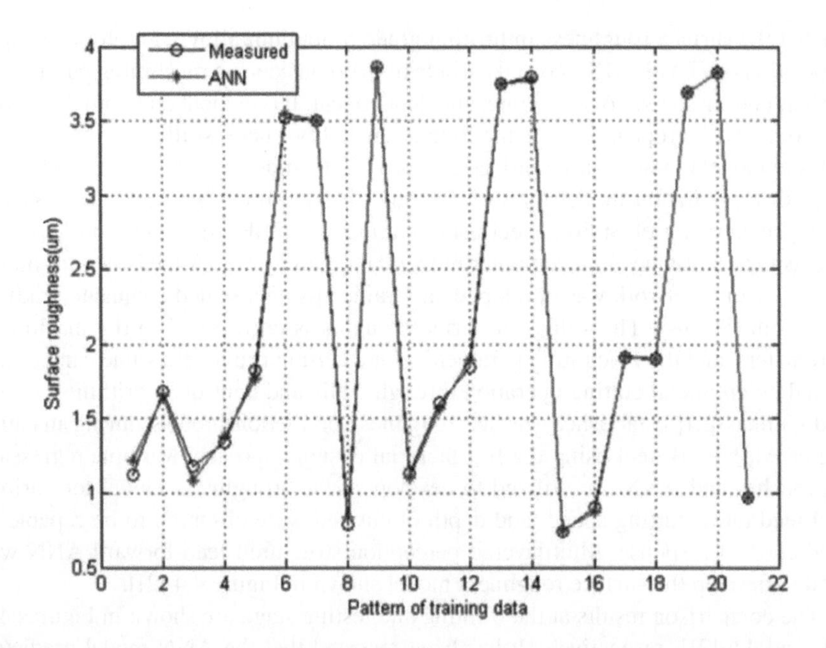

FIGURE 4.5 Comparison of measured and predicted data of surface roughness in the training stage. (From Asiltürk, I. and Çunkaş, M., *Exp. Syst. Appl.*, 38(5), 5826–5832, 2011.)

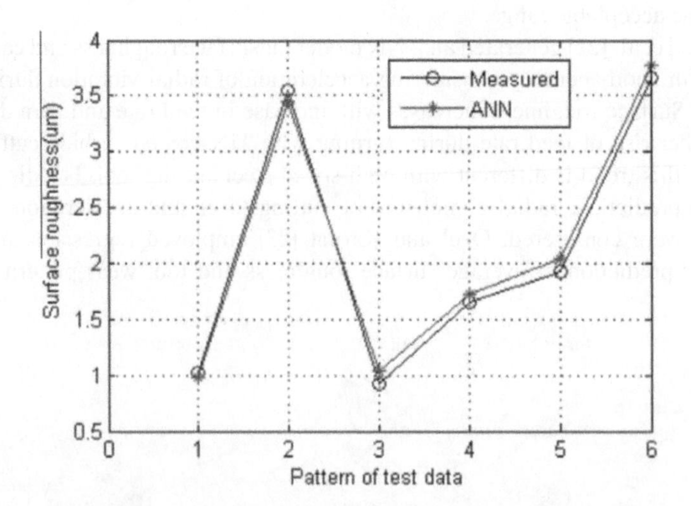

FIGURE 4.6 Comparison of measured and predicted data of surface roughness in the testing stage. (From Asiltürk, I. and Çunkaş, M., *Exp. Syst. Appl.*, 38(5), 5826–5832, 2011.)

hard turning operations. ANN was found to be a better prediction model for surface roughness compared to the regression model. Surface roughness decreases with decrease in feed rate and induces higher tool wear. With increase in cutting speed, tool wear increases but surface roughness decreases. With increase in workpiece hardness, surface roughness decreases but tool wear increases. The performance of a cubic boron nitride (CBN) cutting tool with honed geometry was good with respect to surface roughness and tool wear during hard turning. Sahu et al. [24] established a predictive surface roughness model for grade 5 titanium alloy machining with CBN inserts utilizing cutting parameters along with in process measured cutting force and acceleration amplitude of vibration. A full factorial design was employed for experimentation varying cutting speed, tool feed rate, and depth of cut, respectively. Furthermore, an ANN model was also generated to predict surface roughness. The outcomes of predictions for improved regression models are compared with ANN. RSM and ANN (average absolute error of 7.5%) were found to predict roughness with more than 90% accuracy. ANN has provided improved prediction over RSM models. Basak et al. [25] studied the optimization of hard turning of D2 steel operation with ceramic insert using a radial basis function neural network model. This model is fitted for the prediction of responses such as tool wear and surface roughness with respect to cutting speed, feed rate, and machining time, and trained models are utilized in the optimization. The objective of optimization is minimization of production time and machining cost. The optimization result is dependent on tool changing time and the ratio of operating cost to tool changing cost. Ozel et al. [26] investigated finish hard machining of D2 steel using multi-radii ceramic wiper insert with a developed multiple linear regression model and neural network model for surface roughness and tool wear and compared with non-training data. The neural network model was found to be effective in the prediction of responses for a wide range of cutting conditions. Forces, power, and specific measured forces are taken into consideration for training the algorithm in a neural network. Phate and Toney [27] developed prediction models for responses (material removal rate and surface roughness) during wire electrical discharge machining (WEDM) of an Al/SiCp metal matrix composite (MMC) using dimensional analysis (DA) and ANN techniques (Levenberg–Marquardt back propagation training algorithm) to correlate input parameters such as pulse-on time, pulse-off time, wire feed rate, current, voltage, thermal conductivity of work material, coefficient of thermal expansion, density, and wire tension. An ANN algorithm consists of training phase, testing phase, and validation phase. The structure of an ANN is shown in Figure 4.7 [27].

The comparison results between experimental and predicted values for surface roughness are shown in Figure 4.8 [27], which shows a good fit. Both models are found to be accurate, reliable, and adequate as R^2 values (correlation coefficient) are higher, close to 0.9999 in the training, testing, and validation phases with a lower MSE. However, the ANN model was found to be more accurate and superior compared to the DA model.

Ozden et al. [28] developed an ANN model for cutting force during machining of nonreinforced and reinforced polymide (PA) with a PCD cutting tool. The ANN model was found to be effective in the prediction of cutting force and was in good agreement with experimental results. Karayel [29] developed a feed-forward multilayered neural network model trained using the scaled conjugate gradient algorithm (SCGA), which was a type of back propagation. The machining was conducted for St 50.2 steel with variation in depth of cut, feed rate, and cutting speed for surface roughness parameters (Ra, Rz, and Rmax). The feed rate was the more influencing parameter for surface roughness, which increased rapidly with the rise in feed rate. A

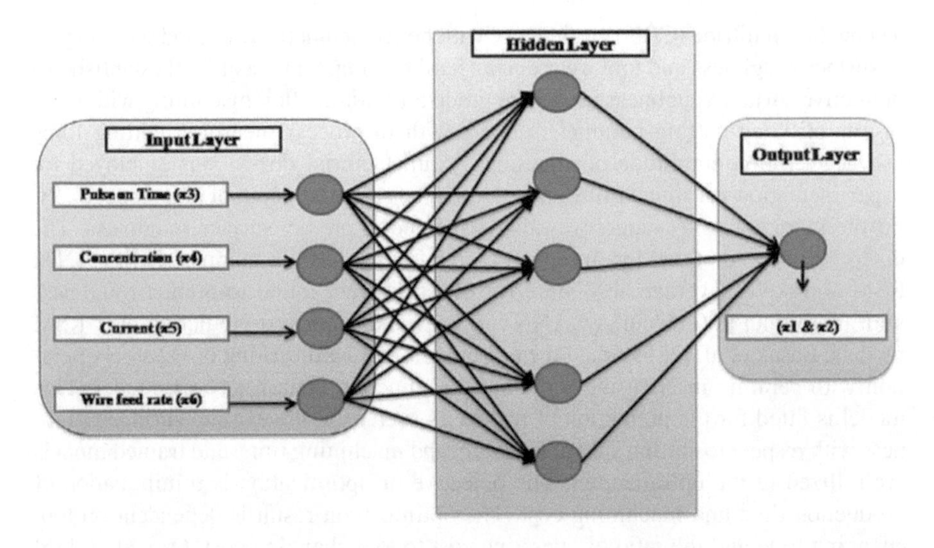

FIGURE 4.7 ANN basic structure. (From Phate, M.R. and Toney, S.B., *Eng. Sci. Technol., an Int. J.*, 22(2), 468–476, 2019.)

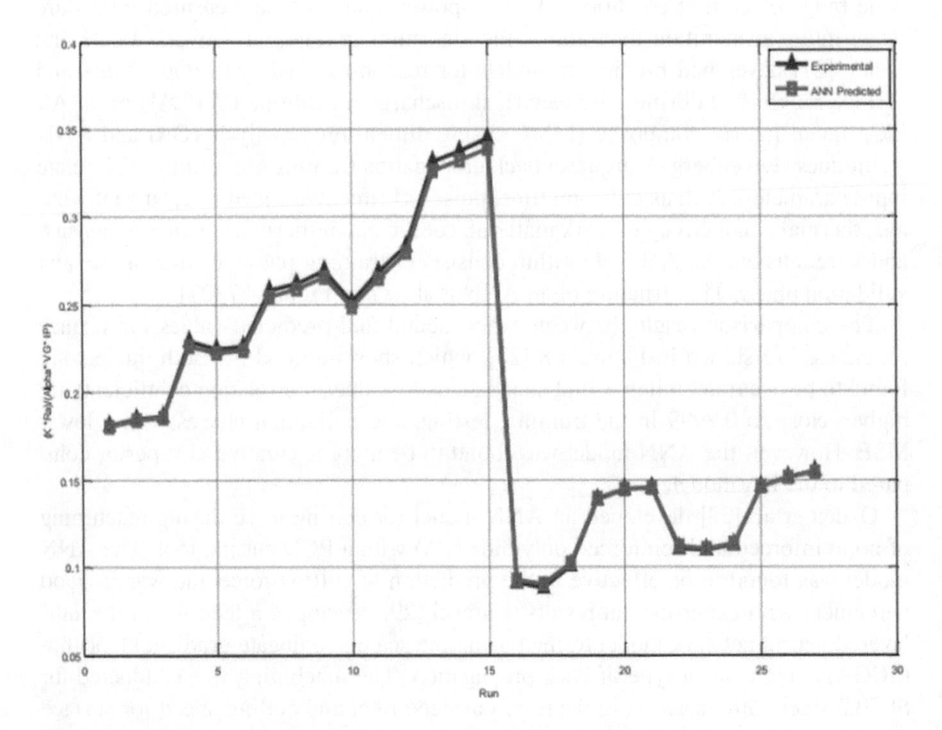

FIGURE 4.8 Comparison of experimental and ANN-predicted values of Ra. (From Phate, M.R. and Toney, S.B., *Eng. Sci. Technol., an Int. J.*, 22(2), 468–476, 2019.)

variable influence has been noticed for depth of cut on surface roughness. The developed ANN model was found to be accurate and effective in relating both input and output variables. Surface roughness can be controlled with the integration of ANN and control system in a CNC machining system. The predicted surface roughness value and the experimentally measured values of surface roughness are very close to each other.

4.2.2 Finite Element Method

Pashaki et al. [30] utilized FEM for the simulation of high-speed dry and cryogenic machining of carbon nanotube (CNT)-reinforced nanocomposites. Dry machining induces higher plastic strain compared to cryogenic machining, and a 12% reduction in plastic strain is observed in cryogenic cooling compared to dry machining. Also a reduction in the cutting temperature by 39.5% was noticed for cryogenic machining compared to dry machining. Machining of pure Al yields continuous chip with less damage at the machined surface, whereas damaged surface has been observed during machining 15 wt% CNT-Al nanocomposite (SPE-01). As far as chip morphology is concerned, chip formation is more continuous for CNT composites, and increase in CNT content influences the chip size. From chip observations, they were found to be conical at a lower cutting speed but at higher cutting speeds, chips become a more open type as observed from three types of nanocomposites such as SPE-01 (15 wt% of CNT), SPE-02 (15 wt% of CNT), and SPE-03 (15 wt% of CNT). CNTs also influence machined surface morphology. The density of the interfacial failure increases with the increase of CNT content and consequently increases cutting force and deteriorates machined surface. Thus FEM can be effectively implemented to investigate the influence of weight fraction, orientation, and length of CNTs during the manufacturing of nanocomposites. Teng et al. [31] established an FEM to simulate a micromachining process with the involvement of cutting edge radius for nano-Mg/SiC MMCs. An arbitrary Lagrangian–Eulerian (ALE) formulation is selected to avoid severe distortion of elements, and the model is shown in Figure 4.9 [31].

The simulation involves tool interaction, chip formation, cutting force, and the Von Misses stress, and strain distribution within the workpiece. The FEM was validated by comparing simulated and experimental results of cutting force and chip morphology (thickness of lamellae on the chip from the experiment with the width of highly strained bands from the simulation), as shown in Figures 4.10 [31] and 4.11 [31], respectively, and were observed to be well correlated. Minimum chip thickness was $0.5R$ as per the simulation results. Size effect can also be obtained from Figure 4.10 based on nonlinear reduction of specific cutting force with increasing uncut chip thickness.

Ray et al. [32] investigated the thermal conductivity improvement of marble-powder-filled epoxy composites experimentally, theoretically, and computationally. FEM was used in a computational way and was compared with the experimental results of thermal conductivity values, and a good agreement was observed. EI-Gallab and Sklad [33] developed a cutting tool model by 3D finite element where material properties are incorporated. This model was used in the turning of Al/SiCp MMCs. The predictions of the model are in good agreement with the experimental

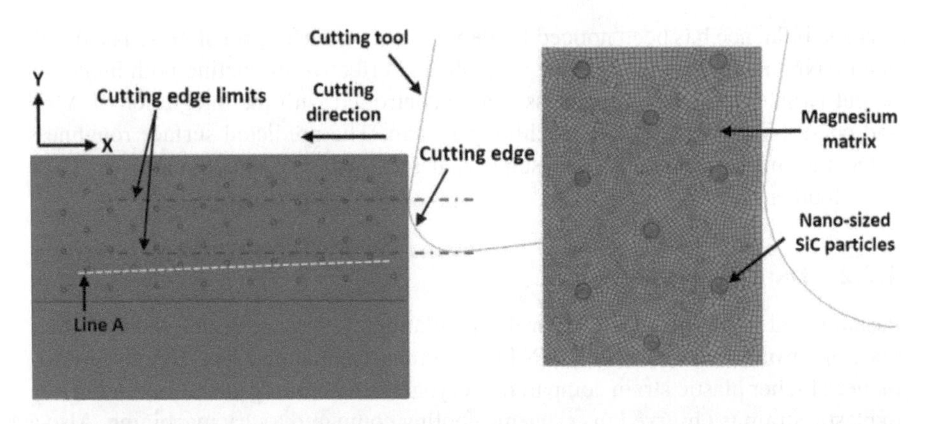

FIGURE 4.9 (a) Schematic representation of the established models for micro-orthogonal machining of Mg/SiC MMCs and (b) local zooming of mesh element. (From Teng, X. et al., *J. Manuf. Process.*, 32, 116–126, 2018.)

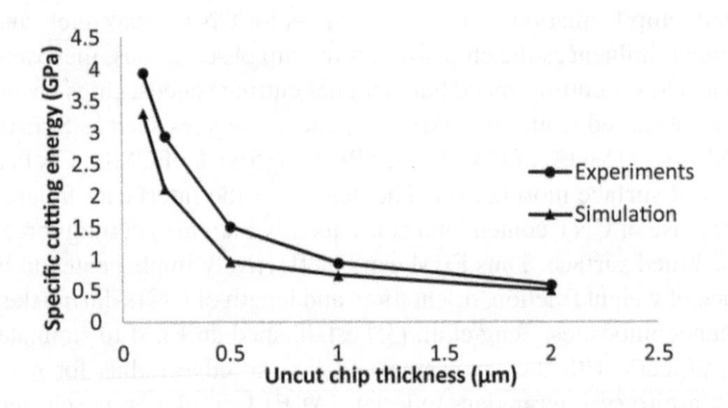

FIGURE 4.10 Specific cutting force at different uncut chip thicknesses. (From Teng, X. et al., *J. Manuf. Process.*, 32, 116–126, 2018.)

measurements and tool wear observations, and was thus found suitable in the selection of cutting parameters, tool geometry, and tool materials, which could lead to reductions in tool wear, machining cost, and tool change downtime. Zhenzhong et al. [34] developed an analytical model of surface roughness of ground-particle-reinforced MMCs. This model was developed considering an undeformed chip thickness model with Rayleigh probability distribution in grinding operation and also considering various removal mechanisms of metal matrix and reinforcement particles. The experimental and predicted results of surface roughness show good accuracy and consistency. Davim and Antonio [35] optimized the cutting conditions in machining of aluminum matrix composites using a numerical and experimental model with a PCD tool. The obtained results show that machining of composite material with PCD tools is perfectly compatible with the ideal cutting conditions and cutting time of the industry and in agreement with optimal machining parameters such as cutting forces, tool wear, and surface finish. Tang et al. [36] investigated dry hard turning

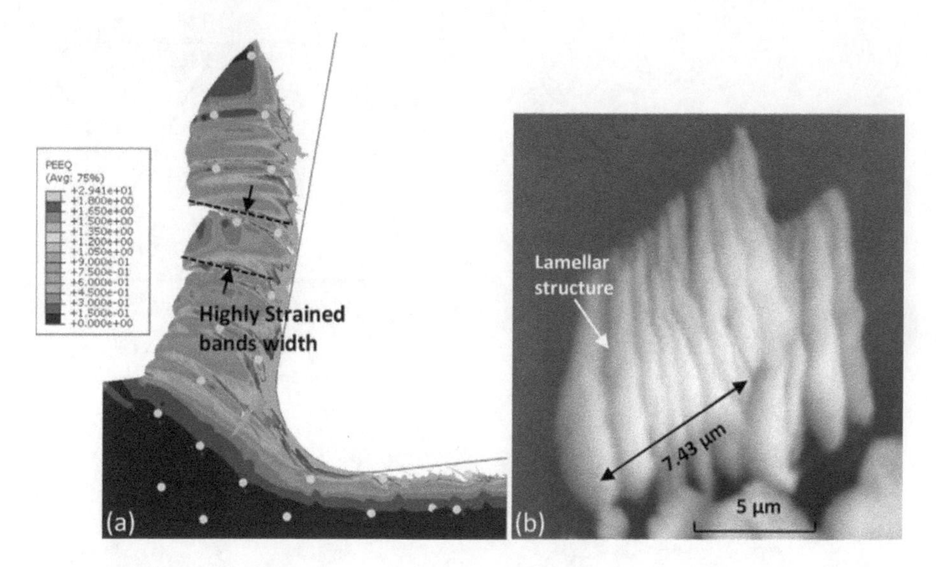

FIGURE 4.11 Chip morphology comparison between (a) simulation and (b) experiment. (From Teng, X. et al., *J. Manuf. Process.*, 32, 116–126, 2018.)

of D2 steel using a CBN cutting tool to study the effect of machining parameters on stress, shear angle, and temperature. The Johnson-Cook model is employed for the development of and FEM to study the effect of cutting speed and depth of cut on responses during hard machining. This model is validated by experimental results. Umer et al. [37] utilized FEM and multi-objective genetic algorithm for the optimization of tool performance during machining aluminum-based MMCs. With variation in cutting speed, feed rate, and reinforcement particle size, cutting forces, chip morphology, cutting temperature, and stress distribution were assessed with this model. Data obtained through Design of experiment (DOE) data are used to find response surfaces using radial basis functions.

4.2.3 Fuzzy Logic

Elsadek et al. [38] developed a fuzzy logic control system to predict surface roughness and tool wear as a function of input variables (cutting speed, feed rate, and depth of cut) and the volume percent of nanoparticulates during dry turning of Al/SiC MMCs and compared with experimental results. They showed a good accuracy of prediction of responses under the selected parametric ranges. Sharma et al. [39] developed a fuzzy-logic-based multi-response prediction model for surface roughness, tool wear, and material removal rate during machining GFRP composites using the Taguchi L27 orthogonal array DOE and optimized multiple performance characteristics using desirability function analysis (DFA). Input parameters such as cutting speed, feed rate and depth of cut are fuzzified into three fuzzy sets – low, medium, and high – as shown in Figure 4.12 [39], whereas responses are fuzzified into eight fuzzy sets – very low, very low, low, medium 1, medium 2, high, very high, and very high – as shown in Figure 4.13 [39]. The developed fuzzy logic model was observed

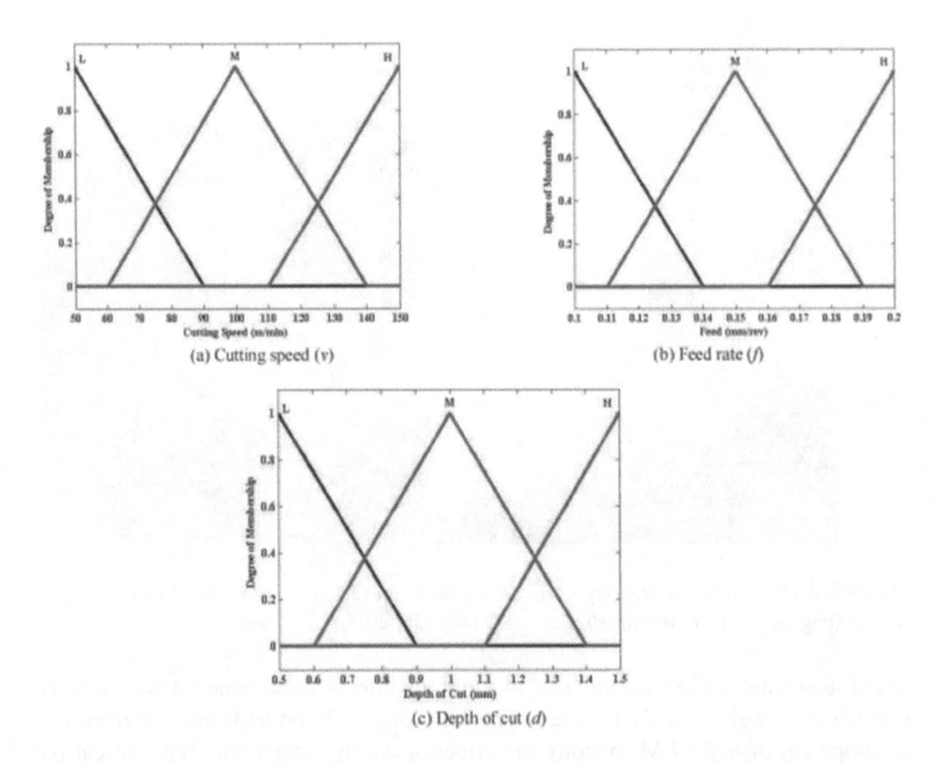

FIGURE 4.12 Fuzzification of input machining parameters. (From Sharma, S. et al., *Proced. Mat. Sci.*, 6, 1805–1814, 2014.)

to be more accurate for the prediction of surface roughness and tool wear as the average error was very less, i.e., 2.74% and 3.06%, respectively. Depth of cut was found to be the most significant parameter on composite desirability.

Jiao et al. [40] utilized a hybrid technique such as the fuzzy adaptive network (FAN) approach for the development of a surface roughness model in the turning operation of 1045 cold-rolled steel with a coated carbide cutting tool. This model was verified using the results of the pilot experiment. Finally, a comparison with the statistical regression was also performed. Abburi and Dixit [41] developed a knowledge-based system that uses fuzzy set theory and neural networks to predict surface roughness when cutting mild steel through carbide and HSS inserts under wet and dry environmental conditions. A massive number of IF-THEN rules are created, which can be minimized to a lesser set of rules by utilizing Boolean operations. It was claimed that the IF-THEN set of rules helped a lot in understanding the behavior of cutting process and the effectiveness of the model. Ramesh et al. [42] employed fuzzy logic to forecast cutting parameters in machining grade 5 titanium alloys under dry conditions. The variables considered were cutting speed, axial feed rate, and cutting depth. A fuzzy-rule-based model was utilized to predict the surface roughness, insert flank wear pattern, and specific turning pressure in the turning process. The effects of the machining parameters on responses have been studied

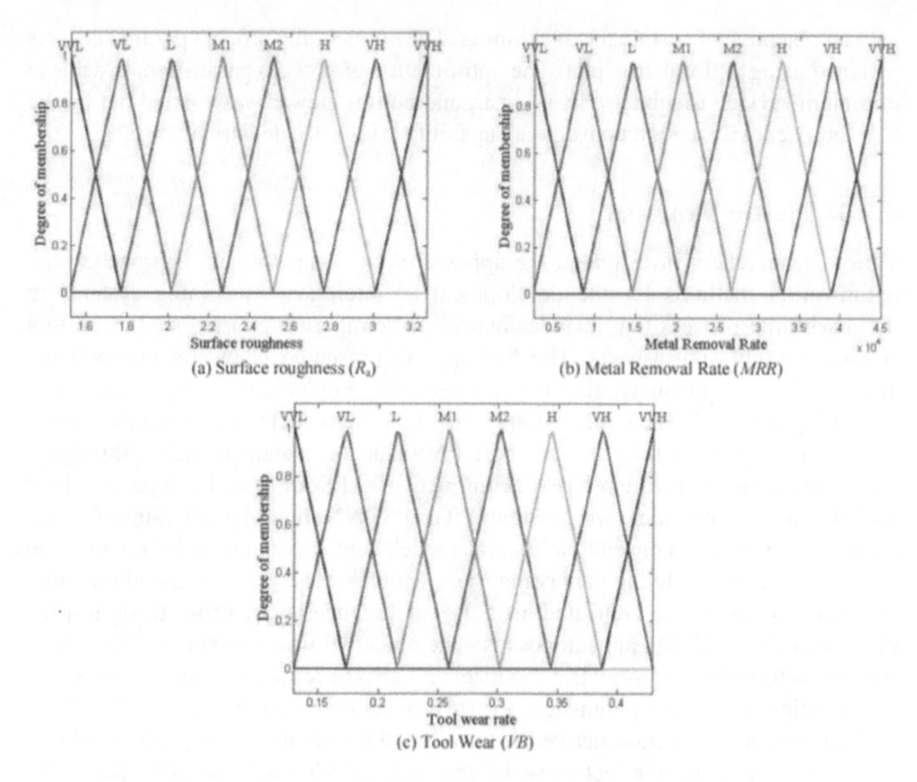

FIGURE 4.13 Fuzzification of output machining parameters. (From Sharma, S. et al., *Proced. Mat. Sci.*, 6, 1805–1814, 2014.)

and also presented. Kirby and Chen [43] established a fuzzy-nets-based supportive average surface roughness forecast model for the machining process utilizing spindle speed, axial feed rate, depth of cut, and tangential vibration signal data. The aim was to build and train a fuzzy-nets-based average surface roughness prediction (FN-SRP) technique that utilizes accelerometer measurements of machining variables and vibration information to predict a turned workpiece's surface roughness. The FN-SRP system was generated using a carbide cutting tool with a CNC slant-bed lathe. Several validation experimental runs show that this technique has a 95% average precision. The sensor and modeling methods chosen in turning operations have excellent potential for both predictive and adaptive monitoring of surface roughness.

4.2.4 Genetic Algorithm

Routara et al. [44] optimized the cutting parameters for surface roughness using the GA approach during CNC machining. Through RSM, a second-order mathematical model has also been developed and validated with F-test and ANOVA. Tamang et al. [45] investigated machinability of Inconel 825 under dry and MQL environment with an objective of sustainable machining. The effect of the process parameters was obtained using a 3D surface plot. The optimization of process parameters to achieve minimum surface roughness, flank wear, and cutting power was carried out by the GA approach with a good convergent capability with a 4% deviation.

4.2.5 Hybrid Modeling

Various researchers investigated the application of computational approaches and optimization methods for the development of intelligent machining systems in the machining process that includes hybrid modeling and reported studies on tool condition monitoring aspects. The findings are presented below for conventional and nonconventional machining process with due emphasis on composites. Yazdi et al. [46] proposed a new methodology for the development of a predictive model for thrust force in drilling operation of PA6-nanoclay nanocomposites through a particle-swarm-optimization-based neural network (PSONN) and compared with a back propagation neural network (BPNN). The PSONN showed good training capacity in comparison to conventional neural models and was found to be reliable and very accurate for modeling nanocomposites. Moghri et al. [47] developed a prediction model through the DOE and artificial intelligence approaches during milling of polyamide-6 (PA-6) nanocomposites. The objective was to optimize the surface roughness to obtain high-quality products and also to reduce machining costs. For optimization of milling parameters for surface roughness, GA is adopted with an explicit nonlinear function derived from ANN for each nanocomposite. Feed rate has the most significant effect on surface roughness followed by spindle speed. The optimal parameters for surface roughness were observed to be the lowest level of feed rate and a middle level of spindle speed. Fathy and Megahed [48] developed ANN and multi-variable regression models for the prediction of abrasive wear rate of Cu-Al_2O_3 nanocomposite materials. ANN was observed to be more effective compared to regression for the prediction of wear rate during analysis of correlation coefficients. Optimization of ANN trained with GA has been compared with ANN trained without GA. In order to find the effect of the factors on the response, sensitivity analysis has also been carried out. Nasr et al. [49] optimized the parameters during milling of graphene-reinforced Ti6Al4V nanocomposites using hybrid adaptive neuro-fuzzy inference system (ANFIS) with the multi-objective particle swarm optimization (MOPSO) method to minimize the feed force, depth force, and surface roughness. A full factorial DOE was used for machining, and ANFIS was used to develop the prediction models and was found to be accurate in the prediction of responses. Compared to the desirability approach, ANFIS-MOPSO has shown better performance in prediction. Krishnamoorthy et al. [50] obtained the optimal parametric combination for multiple characteristics using gray-fuzzy optimization techniques during drilling of CFRP. The fuzzy logic technique has been implemented in order to yield enhanced quality outputs and to reduce data uncertainty. Thus a rule has been written for the fuzzy inference system (FIS) to predict gray fuzzy reasoning grades for all experiments. The membership functions for gray relational coefficients and gray fuzzy reasoning grade are shown in Figure 4.14 [50]. There is an improvement in gray fuzzy reasoning grade over gray relational grade (GRG), as noticed in

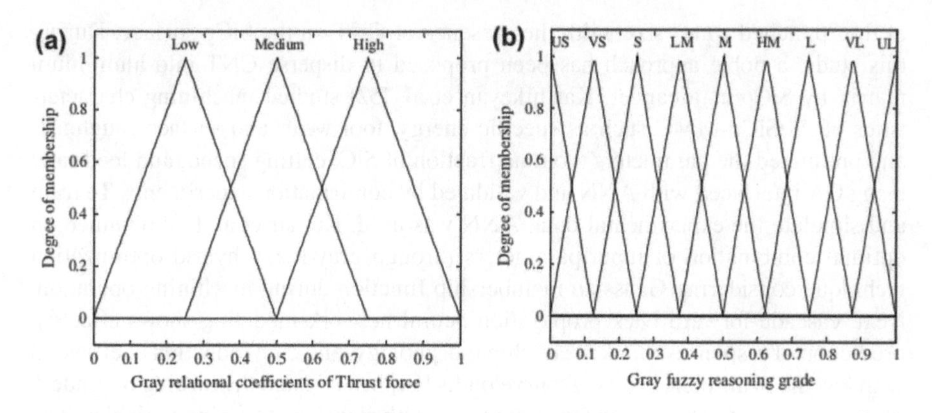

FIGURE 4.14 Membership functions for gray relational coefficients and gray fuzzy reasoning grade. (a) A typical MF for gray relational coefficients and (b) MF for gray fuzzy reasoning grade. (From Krishnamoorthy, A. et al., *Measurement*, 45(5), 1286–1296, 2012.)

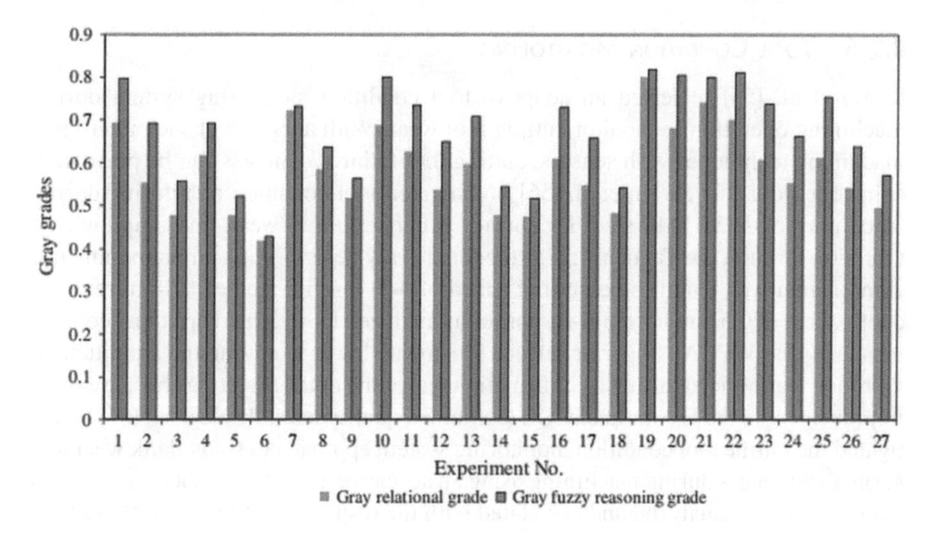

FIGURE 4.15 Comparison of gray relational grade and gray fuzzy reasoning grade. (From Krishnamoorthy, A. et al., *Measurement*, 45(5), 1286–1296, 2012.)

Figure 4.15 [50]. The optimal parameters are thus obtained from gray fuzzy reasoning grade, and the hybrid gray-fuzzy reasoning technique has been found to be useful in the optimization of multiple responses. ANOVA has been employed to study the significance of drilling parameters for responses, and feed rate was found to be the most significant parameter for thrust force, torque, entry and exit delamination factors, and eccentricity.

Li et al. [51] studied the fabrication and mechanical behavior of SiCp particles for CNTs which was reinforced by 6061 Al MMC. The improvement of tensile property has been found for SiC size of 7microns. This may be attributed due to the increase of the "punched zone" size wiith the presence of CNT on the SiCp surface. During this study, a noble approach has been proposed to disperse CNT into aluminium matrix by SiCp as a carrier. Karthikeyan et al. [52] studied machining characteristics of Al/SiCp-MMC such as specific energy, tool wear, and surface roughness and optimized the parameters (volume fraction of SiC, cutting speed, and feed rate) using GA interfaced with ANN and validated by confirmation experiments. To train and simulate the experimental data, ANN was used. Kumar et al. [53] obtained an optimal combination of input parameters through gray-fuzzy hybrid optimization technique considering Gaussian membership function during machining operation. Next, cascade-forward back propagation neural network modeling shows effective prediction of responses (flank wear, chip morphology, and chip reduction coefficient) with low mean absolute errors. Tsourveloudis [54] investigated the turning of grade-5 titanium alloy and studied surface roughness with significant turning variables. The results were transformed into polynomial models through RSM and also through the ANFIS. It was clearly noted that the ANFIS predicts surface roughness with less error under dry and lubricated cutting conditions.

4.2.6 Tool Condition Monitoring

Swain et al. [55] reviewed an adaptive tool condition monitoring system during machining operation to monitor cutting tool wear. With adoption of such intelligent machining techniques with sensors, cutting tool failures/damages can be prevented, enhancing tool life. Swain et al. [56] investigated tool condition monitoring during machining of AISI 1040 steel for surface roughness, tool wear, and amplitude of vibration through the Taguchi L27 orthogonal array design. The optimal parametric combination was found to be cutting speed of 300 m/min, feed of 0.06 m/rev, and depth of cut of 0.5 mm for multi-responses as analyzed by weighted principal component analysis (WPCA). Superior surface finish with lower tool wear and amplitude of vibration was noticed during a confirmation test of the optimal settings because of an improvement in the S/N ratio of CQL (combined quality loss). Panda et al. [57] investigated the online tool condition monitoring system approach to study flank wear and surface roughness during machining using an accelerometer sensor. Vibration signals from sensors are analyzed and correlated with the responses. Mathematical multiple linear and quadratic regression models are developed considering cutting parameters and vibration signals after evaluating the inputs and outputs through the Pearson correlation coefficient. Models are validated and tested for their accuracy. Cutting speed is found to be the most influencing parameter for vibration signal. Panda et al. [58] studied online condition monitoring of tool wear and surface roughness during hard machining through vibration signals generated from accelerometers/sensors. Cutting speed is found to be the dominant parameter for the acceleration amplitude of vibration. Depth of cut, cutting speed, and acceleration amplitude of vibration signals are observed to be strong correlations for flank wear. Cutting speed, feed rate, and acceleration amplitude of vibration signals were observed to be strong correlations for surface roughness. Prediction models (multiple linear regression models) for responses such as surface roughness and flank wear were found to be significant considering the effect of cutting parameters and vibration signals as percentage of error is quite low. Swain et al. [59] studied tool vibration effect and surface roughness during machining through the Taguchi DOE and optimized the process parameters. GRA has been employed to optimize multi-responses such as vibration amplitude and surface roughness simultaneously. Upadhyay et al. [60] implemented the use of amplitude of vibration for the prediction of surface roughness when machining grade 5 titanium alloy and measured through accelerometer, as shown in Figure 4.16 [60].

First, a surface roughness model was developed through first- and second-order regression analysis considering acceleration amplitude of vibration only, and was found to be inaccurate for prediction. Second, considering the association of input parameters and vibration signals on surface roughness through Pearson correlation analysis, a refined multiple regression model for surface roughness was developed considering both machining parameters and vibration signals, and was observed to be accurate as percentage of error was low (4.27%). Finally, an ANN model was also developed and was found to be accurate in the prediction of surface roughness, as the percentage of error was only 4.11%. This ANN model was designed and trained with Levenberg–Marquardt learning rule and the architecture is shown in Figure 4.17 [60].

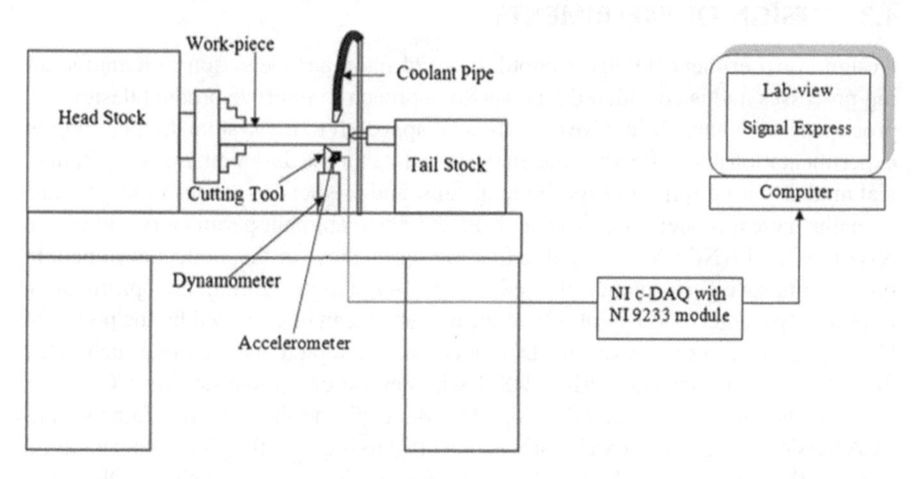

FIGURE 4.16 Schematic of an experimental setup. (From Upadhyay, V. et al., *Measurement*, 46(1), 154–160, 2013.)

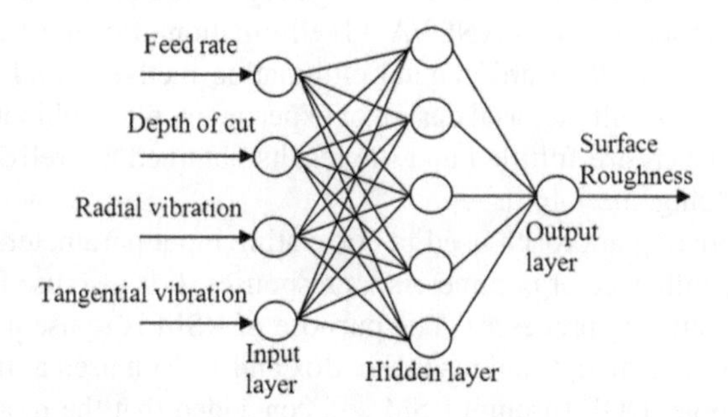

FIGURE 4.17 Neural network architecture. (From Upadhyay, V. et al., *Measurement*, 46(1), 154–160, 2013.)

Lee and Tarng [61] suggested a computer vision approach using a tungsten carbide tool to inspect S45C steel bar's surface roughness. The workpiece surface image was captured through a digital camera. A polynomial network was implemented utilizing an adaptive modeling method of self-organization to develop the relationship among the surface topography aspect and the real, attainable surface roughness in a difference in machining operation. Consequently, the machined component's supportive surface roughness can be foreseen with fair precision if the sketch of the machined surface and machining situation is known. Dharan and Won [62] studied that composite laminates are prone to damage during machining, particularly drilling operations, due to poor transverse strength and low delamination fracture toughness. Thus it is essential to monitor process variables to minimize damage, and the availability of appropriate models with intelligent control schemes would be better for machining such materials. Keeping this in view, experiments were conducted to find the key process parameters for various cutting conditions and an intelligent machining scheme was established.

4.3 DESIGN OF EXPERIMENTS

Design of experiment (DOE) is popularly used for experimentation in manufacturing processes and is considered a powerful approach to improve product design and process performance. It involves a statistical approach to the systematic planning of experimentation and collection and analysis of data for the development of mathematical models with minimum experimental runs, and predicts responses in machining. A mathematical model implies the correlation between input parameters and output responses, and ANOVA is studied to find the significance of the model developed. In order to improve performance and reduce cost in metal machining, the optimization of process parameters is essentially required, which can be achieved by the powerful DOE technique. There are various DOE techniques adopted in machining such as the Taguchi method, factorial design, RSM with central composite design (CCD), etc. Experimental design was first developed by R. A. Fischer in 1920 and data analysis by ANOVA during his research with an objective to improve the yield of agricultural crops [63]. The steps for the development of a prediction model using DOE are as follows: (a) conducting experiment and collection of data as per the design matrix, (b) development of a prediction model through regression analysis, (c) checking the adequacy of the model through ANOVA, (d) elimination of insignificant terms from the model by backward/forward/manual elimination methods, and (e) refining the prediction model. For full factorial design of experiment, all combinations of process parameters and levels are fulfilled and the results obtained are reliable, as reported by Myers and Montgomery [64].

RSM is a modeling approach used in correlating input parameters and responses and to study the influence of parameters on responses. It is also used for the optimization of manufacturing processes. The purpose of RSM is to use a set of designed experiments to obtain an optimal response. Box and Wilson used a first-degree polynomial model to get DOE through RSM and concluded that the model was only an approximation. The response surface regression model is a mixture of statistical and mathematical methods for modeling and analyzing the relationship between input and responses, which can be represented graphically. RSM may fit linear or quadratic equations to analyze the outputs and then identify optimal settings. CCD, also called Box–Wilson design, is adequate for full quadratic models observed in RSM. The geometry of CCDs are classified into faced, inscribed, and circumscribed. CCD has enough design points to estimate the $(n+2)(n+1)/2$ coefficients in a full quadratic model with n factors. Box–Behnken designs are rotatable and require less runs compared to CCDs and are used to calibrate full quadratic models [64].

The Taguchi technique is a popular and widely accepted DOE to produce quality products with less cost. Dr. Genichi Taguchi developed an orthogonal array design with experimental combinations, which makes it more effective to study all the parameters using less number of experiments. Taguchi recommends the use of loss function to measure the deviation between experimental and desired values. Then the loss function is transformed to signal-to-noise (S/N) ratio. There are three types of analysis of the S/N ratio: lower-the-better, higher-the-better, and nominal-the-better. A larger S/N ratio corresponds to optimal levels of the process parameters irrespective of the performance characteristics [65]. Furthermore, the main effects plot gives the trend of the effect of each parameter on response but ANOVA; a statistical approach presents the significant effect of the parameters on response with the percentage of contribution of individual parameters [66]. Finally, a confirmation run is performed to verify the optimal parameters. Thus, the Taguchi design can be divided into three stages: system design, parameter design, and tolerance design. The parameter design stage plays an important role in Taguchi design as it yields optimal parametric condition for the responses. The steps involved in Taguchi parameter design are as follows [67]:

- Selecting the proper orthogonal array (OA) according to the number of controllable factors
- Running experiments based on the OA
- Analyzing data
- Identifying the optimum condition
- Conducting confirmation runs with the optimal levels of all the parameters

4.3.1 Response Surface Methodology

Various researchers in the recent past investigated the application of design of experiment techniques for prediction models and optimization of responses in machining processes that include response surface methodology (RSM). The findings are presented below in conventional and nonconventional machining processes with due emphasis on composite materials.

Fadare et al. [68] utilized the steepest descent method to study the influence of cutting speed, tool feed rate, and cutting depth over the average surface roughness in high-speed turning of grade 5 titanium alloys under a conventional cooling environment using uncoated carbide inserts. It was observed that the average surface roughness was most impacted by axial feed rate followed by the cutting speed and depth of cut. Priyadarshi and Sharma [69] evaluated the effect of machining parameters on responses (cutting force and surface roughness) during turning of Al-6061-SiC-Gr hybrid nanocomposites through a central composite design of the RSM. Optimal parameters can also be obtained through the response surface optimization technique. Feed rate and depth of cut have a major impact on cutting force and an identical effect on surface roughness. Cutting speed has the principal effect on surface roughness and less effect on cutting force. Gopalakannan and Senthilvelan [70] conducted electrical discharge machining of cast Al/B_4C MMCs through an RSM face centered central composite design. Nanoparticles are reasonably well distributed within an Al matrix. XRD and EDS analysis confirms the presence of nano-B_4C particles in the samples obtained from the cast Al/B_4C nanocomposites. Mathematical models have been developed for MRR (material removal rate), EWR (electrode wear rate), and surface roughness respectively. Pulse current and pulse-on time have a major statistical significance on MRR, EWR, and surface roughness. Optimal parametric settings are also obtained from the investigation for MMCs. Finally, a confirmation test was carried out to compare with experimental results and was found to be acceptable, as error was less; it also agreed well with the predicted optimal settings. Hourmand et al. [71] utilized response RSM to study the effect of electric discharge machining input parameters (voltage, current, pulse-on time, and duty factor) on responses such as MRR, EWR, and microstructure change for Al-Mg_2Si MMCs. Input parameters like pulse-on time and its second-order effects influence parameters for EWR. For MRR, voltage, current, two-level interaction of voltage and current, two-level interaction of current and pulse-on time, and the second-order effect of voltage were found to be the most dominant factors during the investigation. Gong et al. [72] investigated end milling of high density polyethlene (HDPE)-MWCNT polymer nanocomposites with the objective of experimental and modeling analysis for surface roughness varying cutting speed and feed. The results were investigated through response table and response function and also in terms of Deborah number, which yields the viscoelastic behavior of the material studied. Gopalakannan and Senthilvelan [73] investigated electric discharge machining of developed a metal matrix nanocomposite of Al7075 reinforced with 1.5% SiC nanoparticles manufactured through the ultrasonic cavitation method. The experiment was conducted through face-centered CCD of RSM. A mathematical model was developed and the effects of process parameters on multi-responses were analyzed by ANOVA. Pulse current has a more significant effect on responses such as MRR, EWR, and surface roughness. From the response surface in Figures 4.18 [73], 4.19 [73], and 4.20 [73], it is evident that MRR increases with increase in pulse current. EWR increases with increase in pulse current and pulse-on time. Surface roughness increases with increase in pulse current, and increases with voltage up to 50 V and then decreases. The process parameters were optimized by the desirability function approach and the values are voltage 50 V, pulse current 8 A, pulse-on time 8 μs, and pulse-off time 9 μs.

Mohanty et al. [74] investigated nanopowder-mixed dielectric electrical discharge machining of Al/SiCp 12% MMC through rotatable CCD. Mathematical model is developed for tool wear rate, MRR, and surface roughness through RSM for correlations. ANOVA yields the significance of the developed model. Noordin et al. [75] studied the effect of feed rate, side cutting edge angle, and cutting speed on surface roughness and tangential component of cutting force utilizing a multilayer coated carbide cutting tool (TiCN/Al_2O_3/TiN) of CNMG120408-FN and TNMG120408-FN

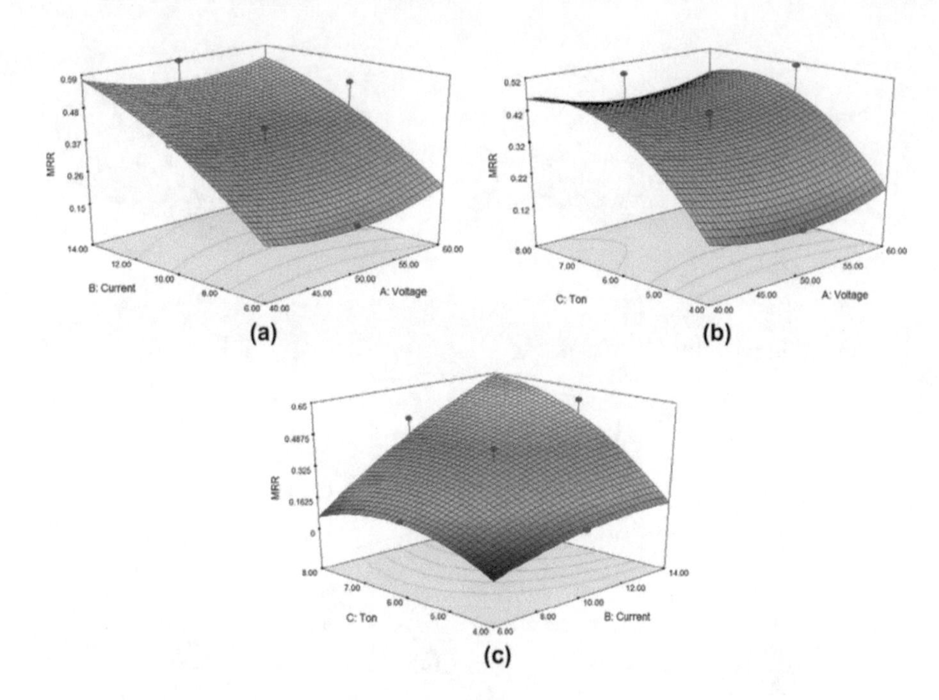

FIGURE 4.18 (a), (b), and (c) show the response graph of MRR of Al 7075 + 1.5wt% SiC. (From Gopalakannan, S. and Senthilvelan, T., *Measurement*, 46(8), 2705–2715, 2013.)

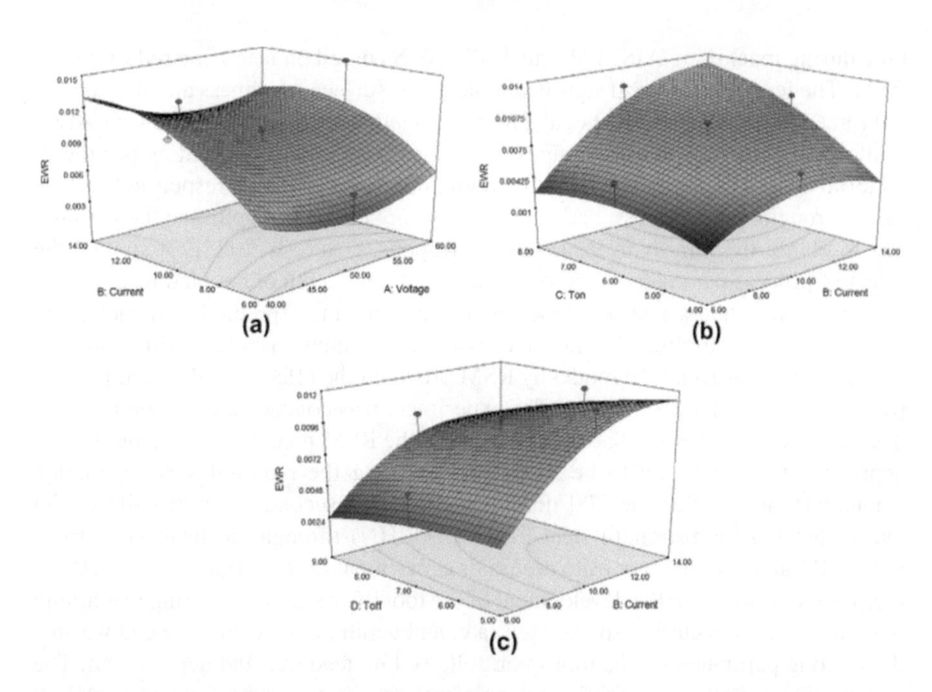

FIGURE 4.19 (a), (b), and (c) show the response graph of EWR of Al 7075 + 1.5wt% SiC. (From Gopalakannan, S. and Senthilvelan, T., *Measurement*, 46(8), 2705–2715, 2013.)

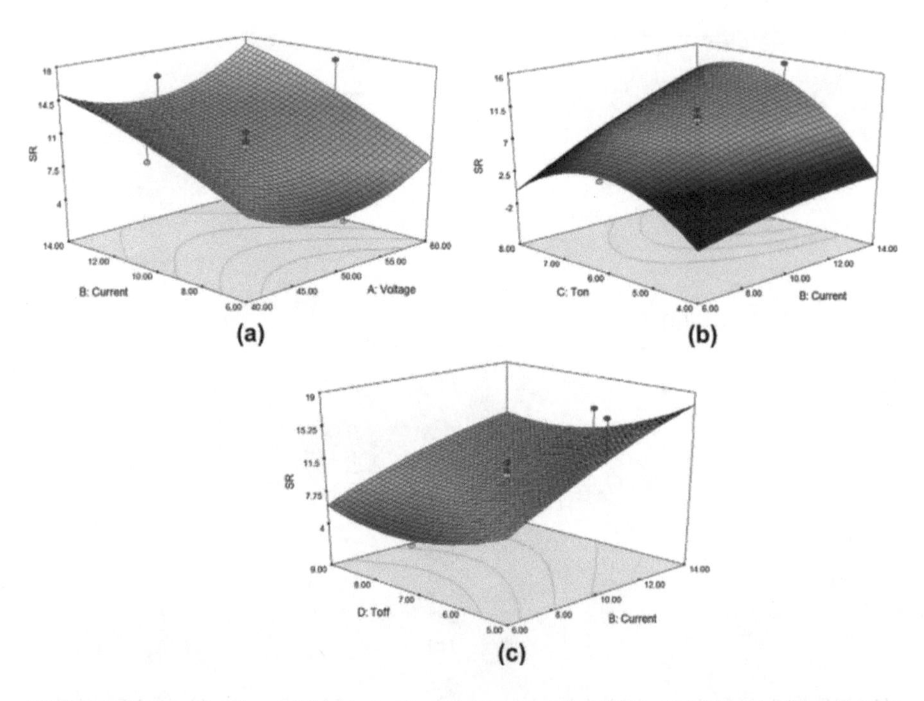

FIGURE 4.20 (a), (b), and (c) show the response graph of SR of Al 7075 + 1.5wt% SiC. (From Gopalakannan, S. and Senthilvelan, T., *Measurement*, 46(8), 2705–2715, 2013.)

type during machining AISI 1045 steel (187 BHN) based on face-centered CCD and RSM. The feed was the most significant factor for surface roughness and the tangential force for ANOVA study. Ic et al. [76] investigated dry machining of SiC or Al_2O_3 reinforced aluminum matrix composites. Experiment was conducted as per the 2^k factorial design method to develop the regression models for the responses, such as surface roughness, hardness, and energy consumption, and to optimize the parameters to obtain the minimum values of the responses. Suresh et al. [77] studied the effect of process parameters (cutting speed, feed rate, depth of cut, and nose radius) on surface roughness and developed a surface roughness prediction model using RSM during machining of mild steel with TiN-coated carbide cutting tools. A second-order mathematical model by RSM provided the effect of individual process parameters on surface roughness. The experiment was conducted based on factorial design of experiment (3^4). The optimization of the RSM model was done by the GA approach and was found to be useful in obtaining the required surface quality. Choudhury and El-Baradie [78] developed first- and second-order tool life model during turning high-strength steel EN24 (290BHN) through an uncoated carbide tool. RSM and central composite design of experiment (2^3 factorial design) methodology were adopted for the development of the tool life model considering machining parameters such as cutting speed, feed rate, and depth of cut. Cutting speed was the dominating parameter on the tool wear, followed by feed rate and depth of cut. The contour plot of tool life determined the optimal cutting condition. Dabnun et al. [79] investigated the machining performance of glass ceramic using uncoated carbide inserts under dry conditions. Applying RSM and factorial design of experiment, a surface roughness model was developed considering machining parameters as cutting speed, feed rate, and depth of cut. As observed from contour plots of surface roughness, feed rate was the major significant parameter that affected surface roughness followed by cutting speed and depth of cut, and also yielded optimal cutting parameters for surface roughness. RSM was found to be an effective approach for collecting a lot of information with a minimum number of runs. Puertas Arbizu and Luis Pérez [80] developed a surface roughness model using RSM (factorial design of experiment) with machining parameters such as cutting speed, feed rate, and depth of cut. The first-order model developed showed about 83.54% variability in Ra and was found to be statistical significant from ANOVA at 99% confidence level, as the p-value was less than 0.01. Feed rate and depth of cut were observed to be the parameters influencing on surface roughness, which increased with increase in feed rate and depth of cut. In order to improve further, a second-order polynomial model was also developed and was found to be statistically significant. Feng [81] applied the fractional factorial design approach to study the effect of turning parameters and their interactions (material hardness, spindle speed, feed rate, depth of cut, and point angle) on surface roughness using multilayer coated carbide inserts ($TiCN/Al_2O_3/TiN$). Feed, nose radius, work material, speeds, and tool point angle were found to be the parameters influencing surface roughness. The most dominant interactions were found between work materials, point angle, and speeds. The depth of cut was found to be an insignificant parameter for surface roughness. The second-order regression model for surface roughness was found to be satisfactory, as evident from the normal probability plot of the residuals and ANOVA analysis. Nalbant et al. [82] experimentally investigated machining of AISI 1030 steel using coated carbide inserts using full factorial design of experiment considering process parameters such as insert radius, depth of cut, and feed rate on surface roughness. Two models for surface roughness were developed and compared: multiple regression model and ANN model. The ANN model was observed to have better prediction over the regression model for surface roughness. As difference in the coefficient of correlation values (R^2 values) is very less, both models could be used also for the prediction of response. Feed rate and insert nose radius were the dominating parameters influencing surface roughness during machining operation. Özel et al. [83] conducted machining of AISI 1045 steel using conventional and wiper geometry of carbide insert through factorial design of experiment to study cutting force, cutting power, and surface roughness. Two prediction models were developed for responses: regression model and ANN model. Wiper carbide inserts produced lower surface roughness compared to conventional inserts. The ANN model was found to be effective for the prediction of surface roughness for a wide range of cutting conditions and can be better implemented in intelligent process planning for wiper insert cutting tools. Tool life is better with lower feed rate. A combination of lower feed rate and the highest cutting speed resulted in better surface finish during machining. Thangavel and Selladurai [84] investigated surface roughness during machining EN24 steel using uncoated carbide cutting tool through fractional factorial design of experiment (central composite rotatable factorial design) considering the machining parameters such as cutting speed, feed rate, depth of cut, and nose radius. A regression model for surface roughness was developed using RSM and was found to be adequate through ANOVA, as the R-ratio of the model exceeded the standard tabulated value of 95% confidence level. Tool nose radius, feed rate and their interactions have the significant effect on surface roughness. The ratio of feed force to tangential cutting force was observed to be reliable and effective to monitor surface roughness during machining. Singh and Kumar [85] investigated tool life and surface roughness during machining of En24 steel with TiC-coated carbide cutting tool through second-order rotatable CCD. A multiple regression model was developed through RSM to establish the correlation between machining parameters (cutting speed, feed rate, and depth of cut) and responses like tool life and surface roughness. The developed prediction model was found to be fitted well and adequate for the prediction of surface roughness and tool life within the parametric ranges, as evident from statistical ANOVA, R^2 value (coefficient of determination), and normal probability plot of the residuals. Al-Ahmari [86] developed three machinability models for tool life, cutting force, and surface roughness using multiple linear regression analysis (RA), RSM, and computational neural network (CNN) during machining austenitic AISI 302 steel and compared the models using statistical and hypothesis testing. Machining was performed through factorial design of experiment considering three process parameters such as cutting speed, feed rate, and depth of cut. The CNN model was found to be better in predicting the responses. Also, the RSM model was found to be superior to regression model in predicting tool life and cutting force. However, all three machinability models had statistical significance and goodness of fit. Gaitonde et al. [87] conducted an extensive study on hard machining of AISI D2 steel using ceramic inserts to replace grinding operation. The experiment was conducted as per full factorial design and a second-order mathematical model was also developed to study the effect of depth of cut and machining time on responses such as cutting force, power, specific cutting force, surface roughness, and tool wear. It is evident from the experimental investigation that the ceramic grade of CC650WG wiper insert performed better with respect to surface roughness and tool wear, whereas CC650 insert performed well with in reductions in cutting force, power, and specific cutting force. Lalwani et al. [88] investigated hard machining experiment of MDN 250 steel through coated ceramic insert to study the influence of process parameters on cutting forces and surface roughness. The experiment was conducted as per RSM and CCD sequential approach. First-order linear model is fitted well for cutting forces. Depth of cut was found to be most influencing parameter for feed force. Feed rate and depth of cut contribute more toward thrust force and cutting forces. Second-order quadratic model fitted well for surface roughness with feed rate as the dominating parameter. The normal probability plot of the residuals indicates that residuals lie on a straight line and the surface roughness model is considered to be significant, as seen in Figure 4.21 [88]. The experimental and predicted values are also close to each other for surface roughness, as evident from Figure 4.22 [88]. The curvilinear profile of the contour plot of surface roughness indicates the fitting of the quadratic model, as seen in Figure 4.23 [88]. It can be seen that at lower feed rate (0.04 mm/rev) and higher depth of cut (0.2 mm), good surface finish can be achieved at any level of cutting speed.

Rana et al. [89] optimized the mechanical properties of AA5083 silicon carbide nanocomposites material by the application of DOE. The effect of process parameters

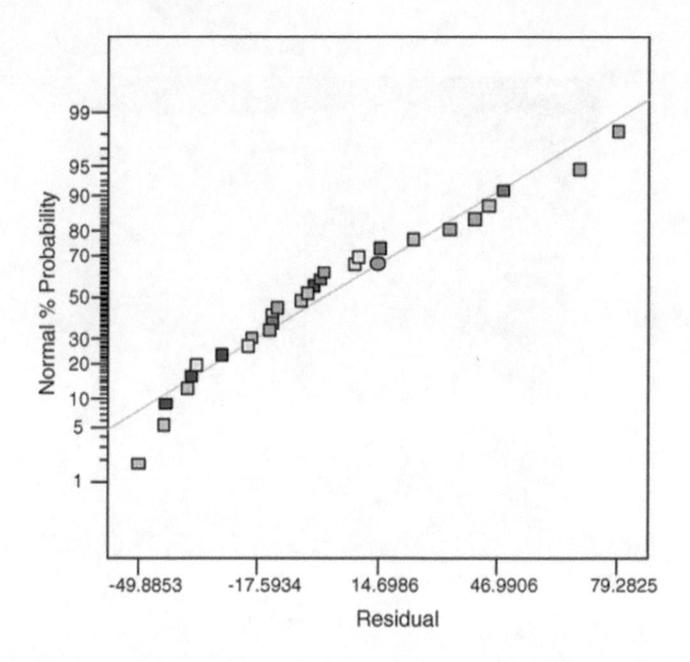

FIGURE 4.21 Normal probability plot of residuals for surface roughness (Ra) data. (From Lalwani, D.I. et al., *J. Mat. Process. Technol.*, 206(1–3), 167–179, 2008.)

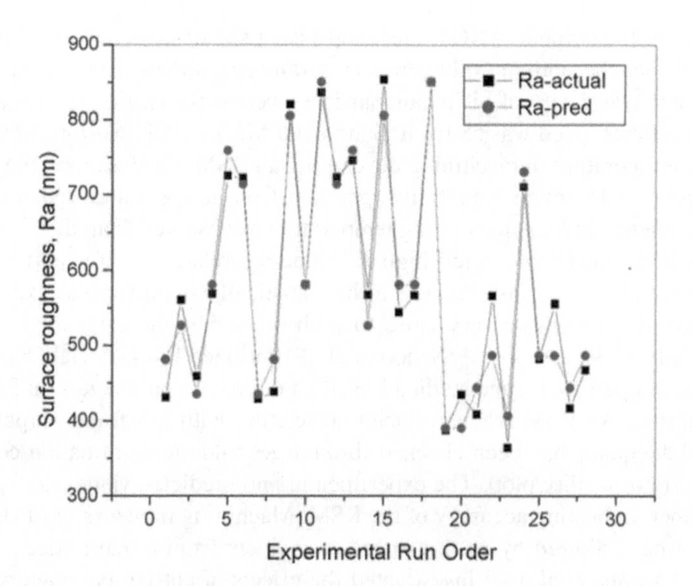

FIGURE 4.22 Actual vs. predicted values of surface roughness (Ra). (From Lalwani, D.I. et al., *J. Mat. Process. Technol.*, 206(1–3), 167–179, 2008.)

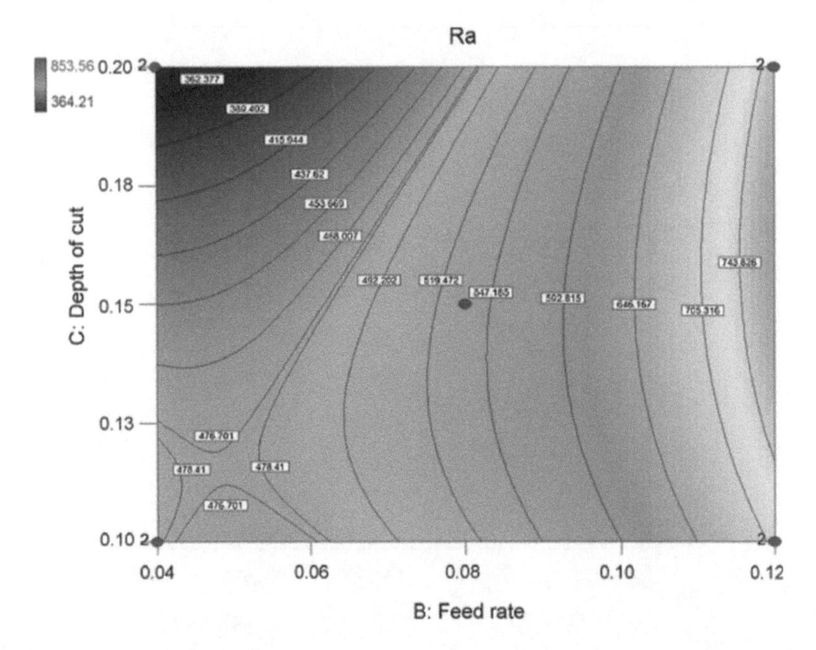

FIGURE 4.23 Surface roughness contour in feed rate and depth of cut plane at a cutting speed of 93m/min. (From Lalwani, D.I. et al., *J. Mat. Process. Technol.*, 206(1–3), 167–179, 2008.)

was tested on the outputs. ANOVA test and Fisher's (F-test) were carried out and it was found that the optimal values of the parameters obtained in the hardness of the specimen was 2 wt% of SiCp nanoparticle whereas the casting temperature was 760°C and stirrer speed was 550 rpm. Sahoo and Mishra [90] investigated the effect of cutting temperature in machining developing an RSM. Furthermore, the parameter was optimized through desirability approach. Cutting speed and feed are found to be the parameters influencing cutting temperature as observed from the main effects plot. The RSM model presented high R^2 value (correlation coefficient) indicating the goodness of fit and was observed to be statistically significant, as experimental and predicted values were very close to each other, i.e., the calculated error was between only 1.88% and 3.19%. Sahoo et al. [91] utilized full factorial DOE to conduct machining operation and studied the effect of process parameters on flank wear at nose corner. An RSM was developed correlating with input and output parameters, and adequacy has been checked through R^2 value (determination coefficient and normal probability plot). The experimental and predicted values are very close to each other, indicating accuracy of the RSM. Machining time was most significant for flank wear followed by cutting speed, as evident from a main effects plot and ANOVA. Czampa et al. [92] investigated the effects of cutting parameters (cutting speed, feed rate, and depth of cut) on surface roughness characteristics (Ra and Rz) during machining of iron-copper-carbon powder metallurgy composites through full factorial DOE. The main effects plot reveals the influence of input parameters on responses. Results reveal that at higher cutting speed, the surface quality of the machined component increases. Sharma et al. [93] investigated wire electric discharge machining (WEDM) of ZrSiO4p/Al6063 MMCs using Box–Behnken design of experiment (BBD) of RSM, and prediction models for responses (cutting rate and surface roughness) were developed. Experimental and predicted results are very close to each other indicating the adequacy of the models. ANOVA and F-test are conducted to identify the significance of WEDM parameters for responses. The desirability concept has been used to obtain the optimal parameters and verified by confirmation experiments. Kumar et al. [94] investigated electric discharge machining (EDM) of Al/10%wt SiCp-MMC to study the influence of electrode shape with size consideration on MRR and EWR using CCD of experiment. Mathematical models for responses are developed through RSM and were found to be adequate and fitted well, as evident from the ANOVA study. The experimental and predicted values are very close to each other because the error was less than 5%. The circular cross section tool shape is observed to be the best tool for higher MRR and lower EWR during EDM. Kumar et al. [95] conducted machining operation of AA6061 composite with carbide cutting tool to yield maximum MRR and minimum surface roughness. The experiment was conducted using the Box–Behnken design of RSM and a mathematical model was developed. The optimized machining parameters are spindle speed of 1800 rpm, feed rate of 0.15 mm/rev, and depth of cut of 1.5 mm for the desired responses. Feed was found to be a major influencing parameter for surface roughness whereas depth of cut influences more on MRR.

4.3.2 Taguchi Design of Experiment

Various researchers in the recent past investigated the application of Taguchi DOE techniques for planning of experimental runs and optimization of responses in the machining process. The findings are presented below for conventional and nonconventional machining process with due emphasis on composite materials.

Hascahk and Caydas [96] utilized a Taguchi-based method to optimize the cutting speed, cutting depth, and axial feed rate for average surface roughness and tool life in cutting grade 5 titanium alloys utilizing CNMG 120408-883 cutting insert. ANOVA and S/N ratio were used to study the output characteristics. The findings showed that the cutting speed and tool feed rates were the main influencing input variables on tool life and surface roughness, respectively. Average surface roughness was related mostly to cutting speed whereas cutting depth had the maximum impact over tool life. Ramana et al. [97] employed Taguchi's robust design methodology for the optimization of process parameters, machining environment, and type of carbide cutting tool during turning of titanium alloy for tool wear. MQL and uncoated carbide tool performed better in the reduction of tool wear as evident from the ANOVA study. Cutting speed has the most significant effect in optimizing the rate of tool wear. Sahoo [98] conducted a machining experiment using the Taguchi DOE considering L27 orthogonal array for surface roughness. S/N ratio and main effects plot were analyzed and the process parameters for surface roughness were optimized. A confirmation run was conducted to verify the results. Furthermore, ANOVA study also investigated the significance of the parameters for response, and feed rate was observed to be significant for surface roughness. Shetty et al. [99] utilized the Taguchi experimental design to optimize the cutting parameters (speed, feed, depth of cut, nozzle diameter, and steam pressure) during turning of MMCs (age-hardened Al6061-15% vol. SiC 25 μm particle size). S/N ratio has been employed to study the effect of cutting parameters on responses (surface roughness, tool wear, and cutting force), and a steam pressure environment was found to be the most significant. Davim [100] investigated the effect of cutting parameters and cutting time on responses such as tool wear, power, and surface roughness during turning of MMCs using the Taguchi orthogonal array DOE. A correlation between input and output parameters was obtained using multiple linear regression analysis. Palanikumar [101] studied the machining characteristics of GFRPs using the Taguchi DOE and RSM for surface roughness. Feed was found to be the most significant parameter for surface roughness followed by cutting speed. The developed RSM model is found to be effective in predicting surface roughness with good accuracy as observed from validation experiments. Davim and Reis [102] studied the machinability aspects of composites (polyetheretherketone reinforced with 30% glass fiber–PEEK GF 30) by PCD and carbide (K20) cutting tools using the Taguchi DOE through L9 orthogonal array. An ANOVA study was also carried out to determine the significant parameters for responses such as power, specific cutting pressure, surface roughness, and international dimensional precision. Cutting velocity has a significance for power and feed rate for specific cutting pressure and surface roughness. Superior performance has been noticed using PCD tool compared to carbide cutting tool during the machinability study. Palanikumar [103] used the Taguchi DOE and Pareto ANOVA approach for the optimization of surface roughness during turning of GFRP composites using a PCD tool. The effect of cutting parameters and their interaction were also studied using S/N ratio and ANOVA. The approach was found to be effective with minimum number of runs. The optimal parameters were obtained from the highest S/N ratio values, i.e., cutting speed (A2:150 m/min), feed rate (B0: 0.1 mm/rev), depth of cut (C2: 1.5 mm). Hussain et al. [104] investigated surface roughness during the machining of GFRP composites using Taguchi DOE and developed a second-order mathematical model by RSM. The RSM model was found to be suitable in the prediction of surface roughness. The parametric influences on surface roughness have also been studied. Satapathy and Patnaik [105] investigated the dry sliding wear behavior of red-mud-filled polyester composites using Taguchi DOE and found the significant control factors affecting the response. Also, an optimal factor combination for minimum wear rate has been achieved. Patnaik et al. [106] Studied the erosion wear performance of GFRP composites using Taguchi DOE, which saves time, material, and costs. It determines the influential factors and their interactions for erosion wear. Finally, the optimal parametric combination for minimum EWR has been achieved. Dakarapu and Nallu [107] optimized the processing parameters of friction stir processing (rotational speed, transverse feed, axial load, and percentage reinforcement) during the fabrication of AA6061/TiB_2 aluminum alloy composites on ultimate tensile strength and hardness of composites using the Taguchi technique. The machining was undertaken as per Taguchi L16 orthogonal array and the results were analyzed through S/N ratio and ANOVA. Rotational speed (65.45%) has a major effect on tensile strength whereas percentage reinforcement (67.3%) on hardness, as evident from ANOVA. The combination parametric level recommended for optimal results for ultimate tensile strength (UTS) and hardness is rotational speed of 1400 rpm, feed of 60 mm/min, and load of 6 N and 4% reinforcement. The optimal results are validated through a confirmation run and are improved. Gray relational analysis (GRA) has been used for optimization of multi-responses and was found to be rotational speed of 1400 rpm, feed of 40 mm/min, load of 7 N and 8% reinforcement respectively and was validated by a confirmation run. Davim [108] investigated turning of free machining steel with carbide cutting tool for surface roughness characteristics such as Ra and Rt using the Taguchi DOE with the L27 orthogonal array. The most significant factors affecting surface roughness were found to be cutting velocity, feed rate, and interaction between cutting velocity and feed as observed from ANOVA. A mathematical model using multiple linear regression/correlation between cutting parameters and roughness was developed and was found to be accurate, as error was less compared to the theoretical model. Singh and Kumar [109] conducted a turning operation of EN 24 steel rod using a TiC-coated carbide insert to investigate feed force by the Taguchi L27 orthogonal array. An optimal combination of machining parameters was found to be cutting speed of 310 m/min, feed of 0.14 mm/rev, and depth of cut of 0.7 mm, and was verified by a confirmation run. Depth of cut (55.15%) and feed rate (23.33%) were observed to be more statistically significant compared to cutting speed (2.63%) for analysis of feed force. The predicted value of feed force was between 149.55 to 189.09 N during the estimation of optimal values at 95% confidence level. Gusri et al. [110] utilized Taguchi DOE of the L27 orthogonal array to conduct a machining experiment on Ti-6Al-4V ELI with coated and uncoated cemented carbide tools varying parameters such as cutting speed, feed rate, depth of cut, and tool type. An optimization of process parameters was carried out by the Taguchi technique considering S/N ratio and a main effects plot for tool life and surface roughness. Cutting speed and type of tool had a strong influence on tool life, whereas for surface roughness, feed rate and type of tool were observed to have a significance, from the ANOVA study. The optimal parametric combination for tool life was cutting speed of 55 m/min, feed rate of 0.15 mm/rev, and depth of cut of 0.1 mm, with an uncoated carbide insert. Similarly for surface roughness, the optimal combination of cutting speed, feed rate, depth of cut, and type of tool was found to be 75 m/min, 0.15 mm/rev, 0.1 mm, and CVD-coated carbide insert, respectively. Dabade and Joshi [111] utilized the Taguchi L27 orthogonal array design in machining Al/SiCp-MMC using CBN inserts with conventional and wiper geometry. The objective is to correlate the quality of surface induced with the type of chips formed in machining MMCs. Gross fracture with higher shear angle occurred during machining of coarser reinforcement composites. Secondary crack formation occurred in machining finer reinforcement composites. Sahoo and Pradhan [112] optimized the process parameters (cutting speed, feed rate, and depth of cut) for surface roughness and flank wear during turning of Al/SiCp MMCs with an uncoated carbide insert using the Taguchi DOE(L9 orthogonal array) and found it to be effective. Lower-the-better characteristics have been used to calculate the S/N ratio, and optimal parameters were selected considering higher mean S/N ratio values and main effects plot, as shown in Figures 4.24 [112] and 4.25 [112], respectively. The optimal parameters for flank wear were cutting speed of 60 m/min, feed rate of 0.05 mm/rev, and depth of cut of 0.4 mm. Similarly, the optimal parameters for surface

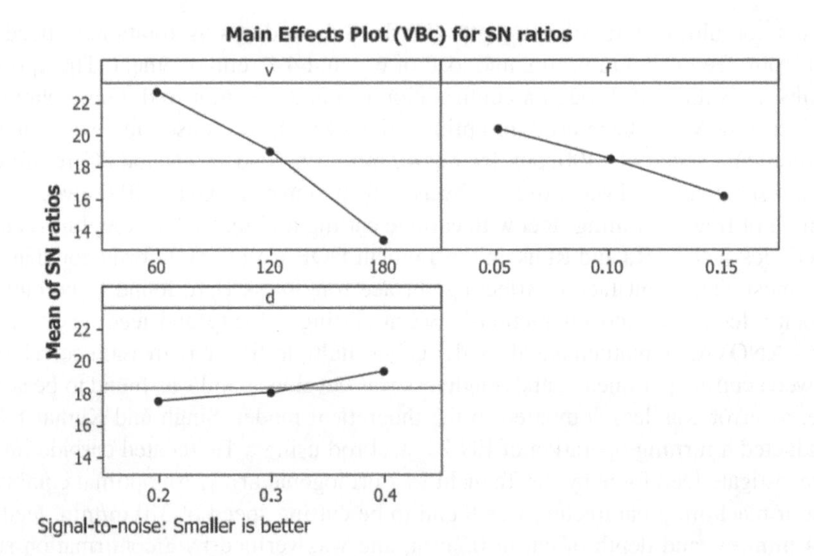

FIGURE 4.24 Main effects plot for flank wear. (From Sahoo, A.K. and Pradhan, S., *Measurement*, 46(9), 3064–3072, 2013.)

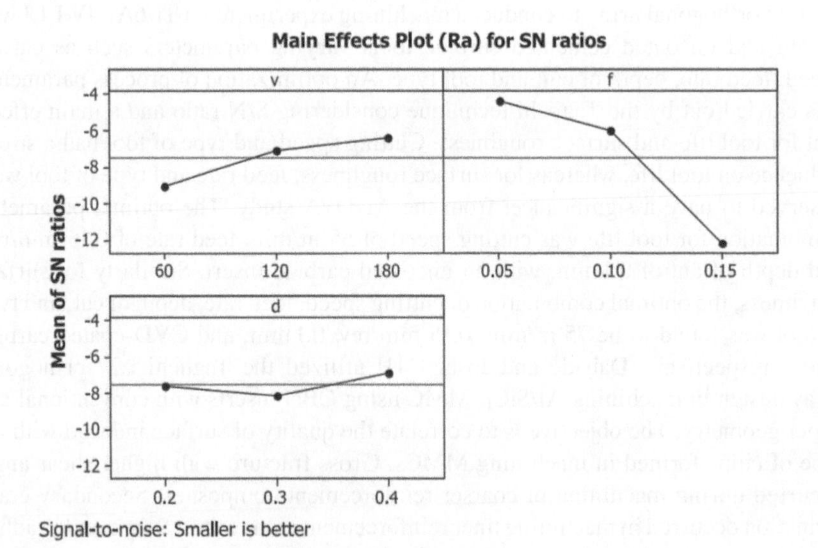

FIGURE 4.25 Main effects plot for surface roughness. (From Sahoo, A.K. and Pradhan, S., *Measurement*, 46(9), 3064–3072, 2013.)

roughness were cutting speed of 180 m/min, feed of 0.05 mm/rev, and depth of cut of 0.4 mm, verified by some confirmation experiments. Mathematical models were also developed through regression analysis and were found to be significant.

Sahoo et al. [113] studied the machinability of Al/SiCp MMCs using TiN-coated carbide cutting tool using the Taguchi L9 orthogonal array DOE for flank wear, surface roughness, and chip morphology. Regression models were also developed with high R^2 values. The significant process parameter for flank wear is cutting speed, and for surface roughness it was feed rate. Parida et al. [114] obtained optimized parameters for surface roughness during machining of GFRP composites with graphite/fly ash filler using the Taguchi L9 orthogonal array DOE. Feed rate was the most significant parameter for surface roughness followed by depth of cut. Rout et al. [115] optimized parameters for erosion wear performance of polyester-GF-granite hybrid composites using Taguchi DOE. Significant parameters affecting wear rate were impact velocity, filler content, impingement angle, and erodent size. Ray et al. [116] conducted erosion wear experiment using Taguchi DOE for polyester composites reinforced with E-glass fiber and filled with different weight proportions of granite and fly ash. The effects of the significant factors were also evaluated. An ANN predictive model was found to be effective with good agreement compared to the mathematical model. A similar experiment has been conducted for glass epoxy composites filled with marble waste using ANN by Ray et al. [117]. Das et al. [118] optimized machining parameters using the Taguchi approach for Al/SiCp MMC with uncoated carbide cutting tool for surface roughness characteristics (Rz and Rt). Using ANOVA, feed rate was found to be the significant parameter for Rz and spindle speed for Rt. Linear regression models are found to be adequate as verified. Tiwari et al. [119] conducted machining experiment using the Taguchi L9 orthogonal array design to investigate output characteristics such as surface roughness, MRR, and chip reduction coefficient. Statistical analysis like ANOVA, surface plot, contour plot, and main effects plot have been analyzed and correlation mathematical models have also been developed through regression theory and were observed to be satisfactory. Mishra et al. [120] conducted a machining experiment of biocompatible titanium alloy using the Taguchi L16 orthogonal array DOE for flank wear and surface roughness. The responses are analyzed by statistical techniques such as ANOVA with F-test and P-value (percentage of contribution), main effects plot, and 3D surface plot. Cutting speed (63.57%) and feed rate (71.48%) were found to be the significant parameters for flank wear and surface roughness, respectively. Sahoo and Sahoo [121] investigated machining operation of D2 steel using the Taguchi L27 orthogonal array DOE to optimize cutting parameters for surface roughness. Feed rate was observed to be the most significant parameter for surface roughness from ANOVA analysis. A mathematical model developed by RSM proved to have accuracy in the prediction of responses and was adequate at a 95% confidence level. Manna and Bhattacharyya [122] optimized cutting parameters for effective turning of Al/SiC-MMC using a fixed rhombic tooling system. An orthogonal L27 array was used for 3^3 factorial designs, and ANOVA was employed to investigate the influence of cutting speed, feed rate, and depth of cut on surface roughness height Ra and Rt. Mathematical models relating to surface roughness height Ra and Rt were established to investigate the influence of cutting parameters during turning of Al/

SiC-MMC. Ramabalan et al. [123] optimized parameters (pulse-on time, pulse-off time, peak current, and wire feed) during WEDM of AA7075/TiB_2 in situ MMCs using the Taguchi L9 orthogonal array DOE, S/N ratio, and ANOVA. The optimal process parameters for MRR have been obtained through the Taguchi optimization technique and were found to be effective from a confirmation experiment. Many investigations have been carried out on erosion wear behavior of polyester hybrid composites, GFRP composites, glass-reinforced polyester-fly ash composites, glass-polyester composites using silicon carbide filling, multiphase hybrid composites consisting of polyester reinforced with E-glass fiber, and ceramic particulates using the Taguchi orthogonal array DOE, and optimal parametric results and models have been determined by Pataniak et al. [124–130].

4.4 MULTI-CRITERION MACHINING OPTIMIZATION TECHNIQUES

Machining is defined as a chip forming metal removal process such as conventional turning, milling, drilling, and grinding with the help of a single- or multi-point cutting using conventional or CNC machine tools. Therefore, the prediction of machining behavior and parametric optimization are current areas of research in metal machining. Selection of appropriate machining parameters is necessary and worthwhile for the improvement in product quality and productivity, and reduction in manufacturing cost in machining/manufacturing industries. Thus the study of optimization techniques plays a crucial role in metal machining for overall productivity. Single-response optimization has so many limitations due to the complex nature of metal machining processes and thus can be avoided by adopting appropriate multi-objective optimization techniques.

Composite materials are popularly used in aerospace and automobile industries because of their properties such as light weight, high modulus, high specific strength, and low thermal expansion. In spite of good properties, machinability of these materials is poor due to their nonhomogeneous and anisotropic properties and it also depends on fiber type, resin type, fiber orientation, and process of manufacture. Thus cutting tools used during composite machining, such as turning, drilling, milling, and in nonconventional machining, suffer a lot of rapid tool wear and failure. The main responses in composite machining investigation are surface roughness, cutting temperature, cutting force, thrust force, metal removal rate, and tool life and wear. The quality of machined components and productivity are two conflicting trends during machining and thus need to be optimized to achieve the desired outcomes. Thus several single- and multi-criteria decision-making optimization techniques are adopted the in machining of composite materials to obtain the desired cutting parameters. The various optimization techniques include the Taguchi quality loss method, TOPSIS (technique for order of preference by similarity to ideal solution), utility concept, AHP (analytic hierarchy process), GA (genetic algorithm), PCA (principal component analysis), PSO (particle swarm optimization), GRA (gray relational analysis), desirability function approach, satisfaction function and distance-based approach, teaching learning based optimization (TLBO), and hybrid optimization techniques, such as back propagation neural network (BPNN)-PSO, Taguchi-GRA, Taguchi-GA,

TOPSIS with GRA fuzzy logic, GRA-fuzzy logic, Taguchi-WPCA (weighted principal component nalysis), whale optimization algorithm (WOA) coupled with GRA, GRA-ANN, Taguchi-TOPSIS, GRA-PCA, Taguchi-GRA-PCA, WASPAS (weighted aggregated sum product assessment), and MOORA (multi-objective optimization based on ratio analysis) [131].

4.4.1 Gray Relational Analysis

Various researchers in the recent past investigated the application of multi-criteria optimization techniques for the optimization of responses in machining processes. The findings are discussed in the conventional and nonconventional machining process with due emphasis on composite materials. Roushan et al. [132] conducted side and face milling operation of GFRP composite at different weightage of performances and optimized milling parameters (rotational speed, feed rate, depth of cut, and cutter speed) for multi-responses (delamination and surface quality) by GRA through the Taguchi L9 orthogonal array DOW. GRA is greatly improved and found to be effective in multi-objective optimization. Rajan et al. [133] fabricated carbon fiber-reinforced aluminum laminates and optimized fabrication parameters (layer thickness, orientation, and matrix) using Taguchi GRA for multi-responses such as flexural strength and impact energy. The fabrication was done through the Taguchi L9 orthogonal array design and hand lay-up technique. Layer thickness and matrix were found to be significant parameters for multi-responses by ANOVA study. Sylajakumari et al. [134] optimized the wear parameters (applied load, sliding speed, and sliding distance) for multiple performance characteristics such as dry sliding wear performance of AA6063/SiC co-continuous composites using Taguchi-GRA. The experiment was performed as per the Taguchi L9 orthogonal array design to record coefficient of wear and specific wear rate. An ANOVA was performed to determine significant parameters and all three wear parameters were observed to be significant. The optimal parameters obtained from GRA were verified through a confirmation experiment and were greatly improved through this approach. Kopparthi et al. [135] studied the effect of fiber layers, resin injection pressure on mechanical properties of E-glass/polyester composites utilizing full factorial design. Taguchi-GRA was employed to optimize the control factors for the responses. Mathematical models were also developed using experimental data for correlation of input and output parameters. Siddhi Jailani et al. [136] optimized the sintering process parameters of Al-Si (12%) alloy/fly ash composite using GRA. Al-Si alloy/fly ash composite was produced using powder metallurgy technique. Taguchi's L9 orthogonal array was used to investigate the sintering process parameters. Experimental results indicate that multi-response characteristics such as density and hardness can be improved effectively through GRA. Palanikumar [137] optimized the drilling parameters of GFRP composites with multi-response GRA with Taguchi method and found it to be effective for performance improvement. A schematic of GRA for multi-optimization is shown in Figure 4.26 [137].

Feed rate is the more dominating parameter for responses such as thrust force, surface roughness, and delamination factor followed by spindle speed, as evaluated from ANOVA and also from the response graph in Figure 4.27 [137]. The optimal

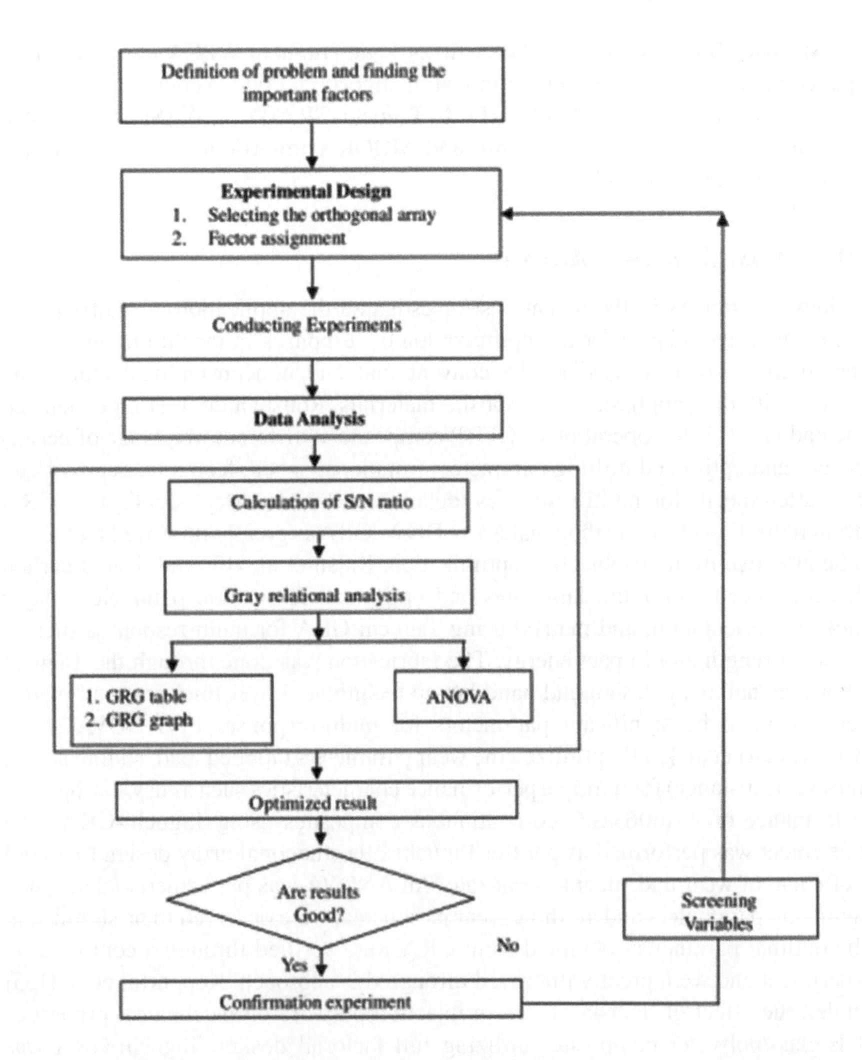

FIGURE 4.26 Optimization procedure used in this work. (From Palanikumar, K., *Measurement*, 44(10), 2138–2148, 2011.)

combination of parameters for multi-responses is determined from the level of highest GRG (Figure 4.28) such as cutting speed at level 4 (2500 rpm) and feed at level 1 (100 mm/min).

Routara et al. [138] optimized process parameters for multi-responses such as MRR and surface roughness using Taguchi-GRA during an abrasive jet machining operation and greatly improved the parameters. Mishra et al. [139] utilized the Taguchi-GRA technique to optimize the input turning variables for output quality characteristics, namely, surface roughness, cutting temperature, and MRR in turning with a carbide tool under dry and spray environments. The optimal results were validated by confirmation run, and an improvement in the GRG was observed. The effect of process parameters (spindle speed, feed rate, depth of cut, and air pressure)

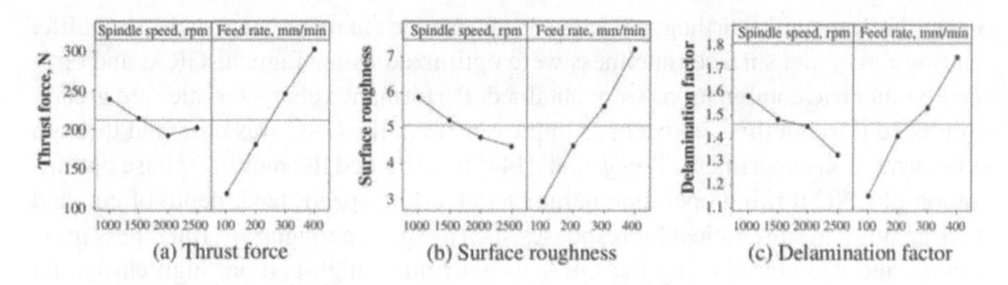

FIGURE 4.27 Response graph for responses. (From Palanikumar, K., *Measurement*, 44(10), 2138–2148, 2011.)

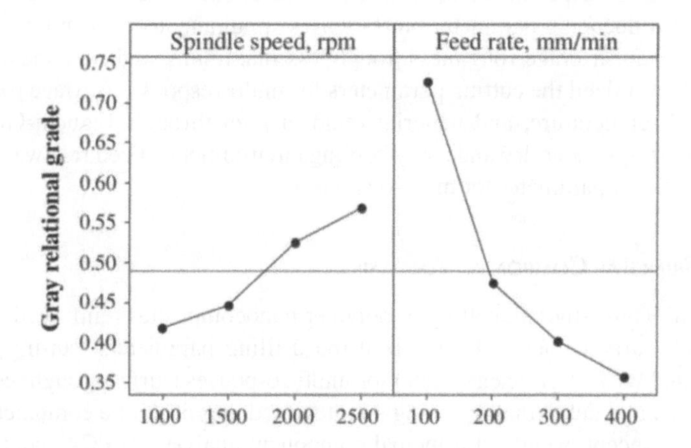

FIGURE 4.28 Gray relational grade graph. (From Palanikumar, K., *Measurement*, 44(10), 2138–2148, 2011.)

on responses was studied by OA and ANOVA techniques. Katamreddy et el. [140] optimized electric discharge machining parameters (peak current, pulse-on time, flushing pressure) for multi-responses such as MRR, TWR, and surface roughness using GRA during machining of Al LM25/AIB_2 functionally graded composite. Peak current had a principal effect on MRR and surface roughness, whereas decreasing trend in the tool wear rate was observed with increase in pulse-on time. Datta et al. [141] developed an RSM model applying full factorial design and optimized the process parameters (current, slag-mix, and basicity index) to obtain acceptable quality characteristics in bead geometry in submerged arc welding process using gray-based Taguchi method and was improved through this approach. Noorul Haq et al. [142] conducted drilling of Al/SiC MMC using TiN-coated twist drill based on the Taguchi L9 orthogonal array design to study responses such as average surface roughness, cutting force, and torque. Multi-response parametric optimization was carried out by GRA and improvement in grey relational grade was observed through a confirmation run. It is evident from the ANOVA study that point angle (43.21%), cutting speed (28.64%), and feed rate (26.21%) have a significant influence on drilling of composites. Lin [143] conducted turning of S45C steel bar by carbide cutting tool using the Taguchi L9 orthogonal array. Performance characteristics such as tool life, cutting force, and surface roughness were optimized using Taguchi-GRA, and optimal parametric combinations were obtained. Performance characteristics are greatly improved through this approach, as improvement in the GRG was observed through a confirmation experiment. Tzeng et al. [144] investigated the multi-response optimization of CNC turning operation parameters (cutting speed, feed, depth of cut, and cutting fluid mixture ratios) for responses such as surface roughness, roughness maximum, and roundness using the GRA in machining high-carbon, high-chromium steel (25 HRC) with TiN-coated carbide insert. The recommended levels for optimal parameters as evident from the average GRG were cutting speed of 155 m/min, feed rate of 0.12 mm/rev, depth of cut of 0.8 mm, and 12% cutting fluid ratio, verified by a confirmation experiment. Operating parameter such as depth was observed to be the most influencing parameter on responses simultaneously when minimization is considered for average roughness, roughness maximum, and roundness. Mishra et al. [145] optimized the cutting parameters for multi-responses (surface roughness, cutting tool temperature, and material removal rate) through Taguchi-GRA (L16 orthogonal array) under dry and spray cooling environments. Feed rate was found to be the significant parameter for multi-responses.

4.4.2 Principal Component Analysis

Kumar et al. [146] studied drilling of polymer nanocomposites reinforced with graphene oxide/carbon fiber and optimized the drilling parameters (cutting velocity, feed rate, and wt% of graphene oxide) for multi-responses (surface roughness, thrust force, torque, and delamination) using an integrated approach of a combined quality loss (CQL) concept, weighted principal component analysis (WPCA), and Taguchi technique. Finally, optimal parameters were validated through a confirmation test. Cutting velocity was found to be the most significant parameter for multi-responses followed by feed rate and wt% of graphene oxide as obtained from an ANOVA study. Swain et al. [147] optimized multi-responses such as flank wear and surface roughness during turning of Al/SiCp nanocomposites using PCA coupled with Taguchi orthogonal array. Significant process parameters were obtained through ANOVA analysis. Cutting speed was found to be the dominant parameter for flank wear, whereas depth of cut and feed rate were for surface roughness. Ukamanal et al. [148] conducted machining experiment under a spray cooling environment utilizing the Taguchi L16 orthogonal array design. Furthermore, the output responses (cutting temperature, tool wear, and surface roughness) were simultaneously optimized by multi-response optimization techniques such as Taguchi-WPCA and verified by a confirmation experiment. Panda et al. [149] investigated hard machining of EN 31 steel through the Taguchi L9 orthogonal array design using coated carbide cutting tool to study surface quality characteristics (Ra, Rz, and Rt). The correlations of machining parameters with responses are modeled by multiple linear regression analysis and were found to be fitted well and accurate because of higher R^2 value. Parametric optimization was carried out using a multi-response optimization technique such as Taguchi-WPCA and resulting in cutting speed of 140 m/min, feed of 0.04 mm/rev, and depth of cut of 0.5 mm. The proposed methodology shows great improvement from confirmatory tests. It is evident from ANOVA that feed rate and depth of cut influenced more on multi-responses. Kumar et al. [150] performed the WPCA technique to obtain optimal process parameters in hard turning of JIS S45C with coated carbide cutting tool and achieved a depth of cut of 0.3 mm, feed of 0.05 mm/rev, and cutting speed of 120 m/min for flank wear, surface roughness, and chip morphology. The S/N of CQL value was increased by 9.3586 (i.e. about 68.3%) from the initial process parameter setting, improved through this approach, which confirmed that the WPCA approach was one of the best optimization techniques that can be implemented in machining works. The BNN model was found to give accurate predictions of responses compared to the regression model or the Elman recurrent neural network model. Liao [151] utilized WPCA for multi-response optimization and observed it to be effective with significant improvement in CQL from the initial settings. Gonela et al. [152] conducted machining operation of AA7075 steel to investigate material removal rate and surface roughness characteristics (Ra, Rq, and Rz) through the Taguchi L9 orthogonal array experimental design. Multi-response optimization of process parameters (speed, feed rate, and depth of cut) was performed by WPCA-CQL method, resulting in speed of 1500 rpm, feed rate of 0.2 mm/rev, and depth of cut of 0.5 mm. Feed rate had a major influence on multi-responses during ANOVA study. Sahoo et al. [153] conducted machining of aluminum alloy using RSM Box–Behnken DOE to study surface roughness and tool vibration. An RSM model was developed taking input parameters (spindle speed, feed and depth of cut) and responses and was found to be adequate from an ANOVA study. Process parameters were optimized by a multi-response optimization technique, i.e., WPCA and an improvement in the multiple response performance index (MPI) was noticed from a confirmation test. Roy et al. [154] studied three zones of cutting temperatures, namely, at chip-tool interface (T), at flank face (Tf), and at machined work surface (Tw) during machining of AISI 4340 steel under a pulsating MQL environment and optimized cutting parameters. As observed, the optimal machining parameters were found to be a cutting depth of 0.1 mm, axial tool feed rate of 0.08 mm/rev, cutting speed of 100 m/min, and pulse rate of 2 s using WPCA along with the Taguchi technique. Linear regression models of cutting temperatures fitted well and were accurate, as absolute mean errors were less than 4%. Cutting speed has a major impact on T and Tf, whereas depth of cut influenced more on Tw. Das et al. [155] investigated dry machining of Al 7075/SiCp MMCs and studied the effect of parameters on surface roughness (Ra) and MRR. Surface roughness of the machined component decreased with increase in spindle speed or decrease in feed rate. MRR increased with increase in all three cutting parameters (spindle speed, feed rate, and depth of cut). For multi-response optimization of cutting parameters, WPCA coupled with the Taguchi technique was used. The enhancement in S/N ratio of CQL was observed to be 12.1144 dB from a confirmation experiment. Pamuji et.al [156] optimized multi-responses (surface roughness and MRR) during machining of ST 60 Tool steel under minimum quantity cooling lubrication environment using the Taguchi-WPCA technique. Cutting speed had a principal effect on multi-responses followed by feed rate, type of coolant, and depth of cut. The optimal process parameters were observed to be a cutting speed of 172.95 m/min, feed rate of 0.053 mm/rev, depth of cut of 0.25 mm, and coolant as vegetable oil for the minimization of surface roughness and maximization of MRR. Jayaraman and Kumar [157] carried out multi-optimization of cutting parameters during dry machining of AISI 6061 T6 aluminum alloy using an uncoated carbide insert with the WPCA technique. It was stated that WPCA worked excellently to estimate the optimal condition based on the multi-response performance indicator (MPI) and a validation test. The significance of the cutting parameters were estimated through an ANOVA study.

4.4.3 Desirability Approach

Rao and Padmanabhan [158] investigated electrochemical machining of Al-Si/B_4C composites fabricated by a stir casting process and optimized the process parameters (applied voltage, electrode feed rate, electrolyte concentration, and percentage of reinforcement) for multiple performances (surface roughness and radial over cut) using RSM. The effect of process parameters and their interaction effects were studied using surface plots. The optimal parameters were feed rate of 1 mm/min, voltage 15.25 V, electrolyte concentration of 13.56 g/lit, and 7.36 wt% of reinforcement of B_4C. The predicted models for responses were found to be statistically significant at 95% confidence level and validated. The experimental and predicted values of responses are very close to each other and thus can be safely utilized for the prediction of responses within the ranges studied. Bhushan [159] investigated machining of 7075 Al alloy-15 wt% SiC (20–40 μm) composite with an objective to yield minimum tool wear and maximum MRR, which were dependent on cutting parameters such as cutting speed, feed rate, depth of cut, and nose radius. The overall desirability function was used for simultaneous optimization of responses and was found to be satisfactory. The optimal cutting parameters for minimum tool wear with maximum MRR were cutting speed of 210 m/min, feed of 0.16 mm/rev, depth of cut of 0.42 mm, and nose radius of 0.4 mm. Khamel et al. [160] developed empirical models on AISI 52100 grade steel with a CBN cutting insert using RSM in hard machining. The effect of machining parameters (cutting speed, feed rate, and depth of cut) on responses (tool life, surface roughness, and cutting forces) was also investigated. Multi-response optimization of process parameters was carried out through the composite desirability approach. The proposed methodology for model development and optimization have been found to be reliable to improve the hard turning process. Feed rate and cutting speed have more influence on surface roughness and tool life. Depth of cut plays a vital role in influence cutting force. Khajuria et al. [161] studied the effect of EDM parameters (current, voltage, pulse-on time, and wt% reinforcement) on responses (MRR, TWR, and SR) using RSM. The experiment was conducted through rotatable CCD of RSM. The optimal parametric conditions were found to be current of 12 A, voltage of 60 V, pulse-on time of 60 μm, and reinforcement of 5 wt% Al_2O_3/AA2014 Al composite. An increasing trend in MRR, TWR, and SR was observed with increase in current and pulse-on time. Horng et al. [162] conducted hard turning of austenitic Hadfield steel using an uncoated ceramic cutting tool to study the effect of process parameters (cutting speed, feed rate, depth of cut, and corner radius) on responses such as flank wear and surface roughness through CCD along with ANOVA. The RSM quadratic model along with the sequential approximation optimization (SAO) technique was used to obtain the optimal set of cutting parameters. The RSM-based wear and roughness models predicted the close fit value to experimental data values at 95% confidence level. There was 9.25% and 8.74% reduction in flank wear and surface roughness, respectively, using optimal parameters. Cutting speed is the major influencing parameter for flank wear. Interaction effect of feed nose radius and cutting speed corner radius influenced more on surface roughness. Bhushan [163] investigated the influence of machining parameters (cutting speed, feed, depth of cut, and nose radius) on responses (power consumption and tool life) during machining of Al alloy SiC particle composite using RSM. Multi-response parametric optimization of responses was performed through a composite desirability approach and the significance of parameters were judged by an ANOVA study. The composite desirability value was 0.902 as evident from Figure 4.29 [163] and adequate for the prediction of process parameters for multi-optimization (minimum power consumption and maximum tool life). From contour plots, i.e., Figures 4.30 and 4.31 [163] at maximum desirability value, the predicted value of power consumption and tool life were 1017 Wh and 6.59 min. Multi-responses are greatly improved during confirmation test and thus considered as an effective approach for multi-optimization. At optimal parametric values, power consumption has been reduced by 13.55% and tool life increased by 22.12% respectively.

4.4.4 Genetic Algorithm

Dhandapani et al [164] optimized multi-response EDM parameters (pulse-on time, pulse-off time, current, voltage, and wt% of MWCNT) during machining aluminum matrix reinforced with boron carbide (B_4C) and MWCNT. The experiment was conducted through CCD, and quadratic regression models were developed for

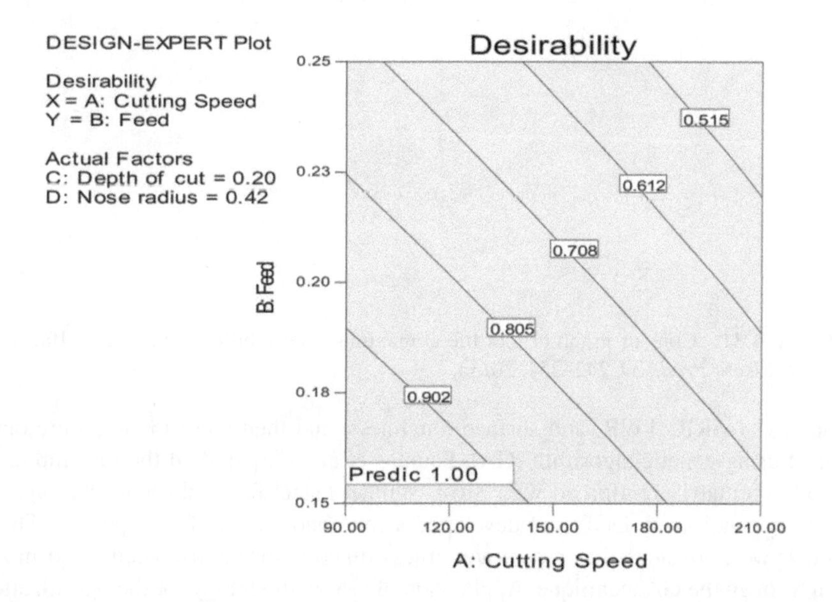

FIGURE 4.29 Composite desirability of power consumption and tool life. (From Bhushan, R.K., *J. Clean. Prod.*, 39, 242–254, 2013.)

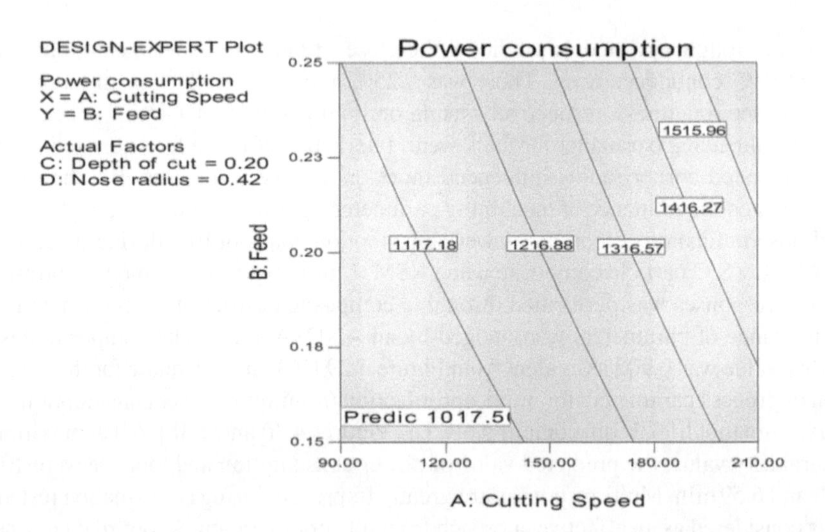

FIGURE 4.30 Contour graph of power consumption at maximum desirability value. (From Bhushan, R.K., *J. Clean. Prod.*, 39, 242–254, 2013.)

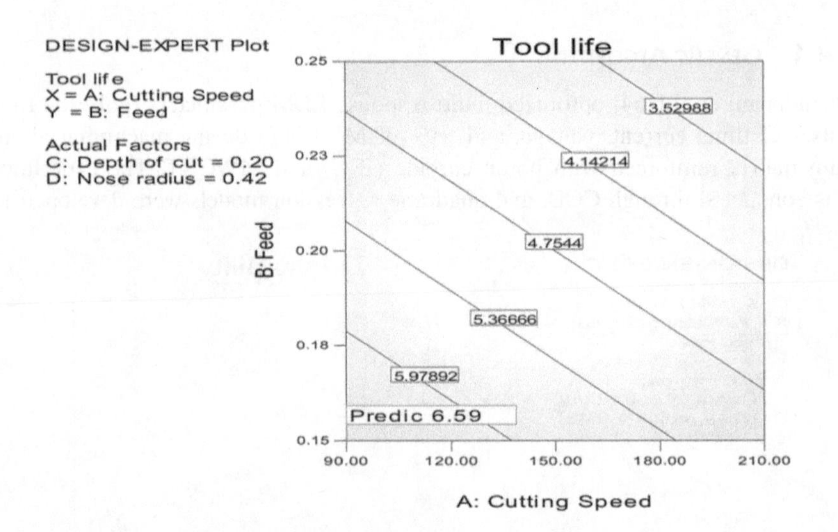

FIGURE 4.31 Contour graph of tool life at maximum desirability value. (From Bhushan, R.K., *J. Clean. Prod.*, 39, 242–254, 2013.)

responses (MRR, TWR, and surface roughness) and then these models were optimized using genetic algorithm (GA). Fountas et al. [165] studied the machining of aluminum matrix reinforced with 316-L stainless steel flakes through the Taguchi L9 orthogonal array DOE and developed a regression model for responses. These models were considered as a common fitness function for multi-objective optimization through the GA technique. Application of GA methodology for the optimization of parameters was found to be effective in machining aluminum matrix particulate composites. Dhandapani et al. [166] studied the electrical discharge machining of aluminum matrix reinforced with boron carbide (B_4C) and MWCNT-reinforced sintered composites. The experiment was performed through CCD and the machining parameters (pulse-on time, pulse-off time, current, voltage and wt% of MWCNT) were optimized through GA for responses (surface roughness, MRR, and TWR). Second-order quadratic models were developed through regression analysis. Kumar et al. [167] investigated machining of unidirectional glass-fiber-reinforced plastics (UD-GFRP) composite through the Taguchi L18 orthogonal array design considering machining parameters such as nose radius, rake angle, feed rate, cutting speed, and cutting environment and measured tangential and feed force. Regression models are developed and the significance of machining parameters was judged through an ANOVA study. The optimization of multi-responses was performed by distance-based Pareto genetic algorithm (DBPGA) and was found to be 39.93 and 22.56 N, respectively, for tangential and feed force, which were very close to the measured values. The procedure involved in DBPGA is shown in Figure 4.32 [167].

4.4.5 Particle Swarm Optimization

Das et al. [168] investigated aluminum-based Al 7075 MMCs reinforced with SiC machining under spray environment using coated carbide insert and measured cutting temperature, surface roughness, and flank wear. Quadratic regression models were developed for responses and were found to be adequate. PSO methodology was used for multi-response optimization of cutting parameters along with individual parametric optimization by the Taguchi technique. PSO outperformed the Taguchi approach. ANOVA was used to judge the significance of cutting parameters for responses. Mishra et al. [169] utilized PSO during machining and analyzed the effect of parameters on responses (surface trait and MRR). The Taguchi L16 orthogonal array experimental design was selected to perform the machining operation. The proposed PSO technique was observed to be efficient for multi-criteria decision-making problems. Soepangkat et al. [170] optimized multi-responses (thrust force, torque, hole entry delamination, hole exit delamination) during drilling of CFRP. The drilling experiment was performed through a full factorial DOE. For modeling the drilling process and to predict the optimum responses, BPNN was used. The optimum combination of process parameters was obtained through PSO methodology and was observed to be accurate, as evident from a confirmation experiment. Response graphs were plotted to study the effect of drilling parameters on responses.

4.4.6 Multi-objective Optimization Based on Ratio Analysis

Biswas et al. [171] studied the effect of input parameters (slurry concentration, feed rate, and power) on responses (MRR, taper angle, and over-cut) during ultrasonic machining of zirconia (ZrO_2) composite using the Taguchi L9 orthogonal array design and optimized the parameters for multi-responses by the MOORA method. Optimized results are verified by a confirmation test. Feed rate influenced more on MRR whereas slurry concentration had a major impact on taper angle and over-cut. Pathapalli et al. [172] carried out machining experiment of stir-casting-fabricated aluminum 6063 as matrix and titanium carbide as reinforcement MMC using the

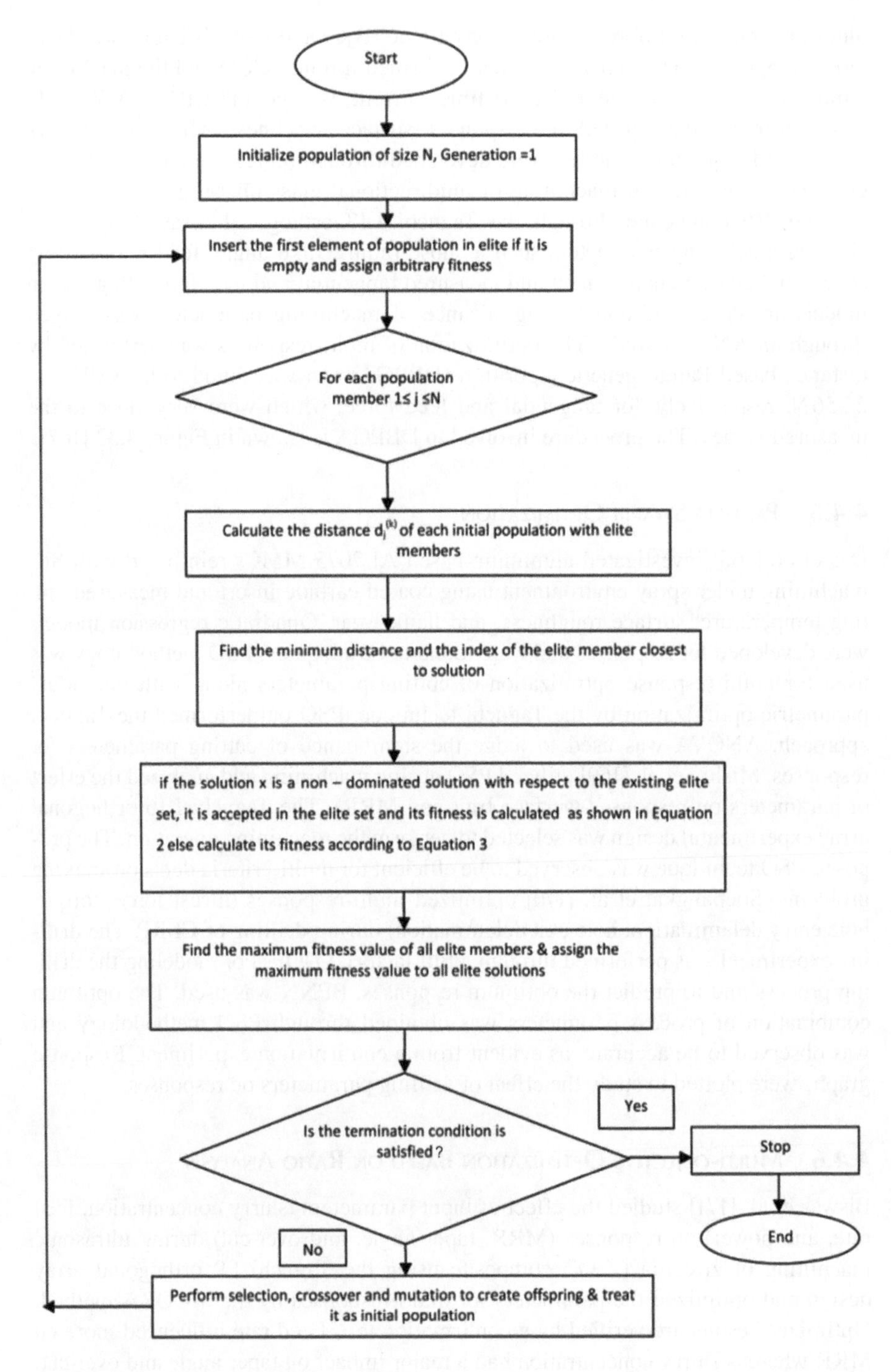

FIGURE 4.32 Flow chart for the DBPGA. (From Kumar, S. et al., *Eng. Sci. Technol., an Int. J.*, 18(4), 680–695, 2015.)

Box–Behnken design. Two multi-criteria decision-making approaches such as MOORA and WASPAS (weighted aggregated sum product assessment) were adopted for multi-optimization of parameters such as cutting speed, feed rate, depth of cut, and reinforcement wt% of composite samples for responses, i.e., MRR, cutting force, and surface roughness. The same optimal parameters settings were observed using both MOORA and WASPAS approaches, which will be immensely helpful for manufacturing industries.

4.4.7 Technique for Order of Preference by Similarity to Ideal Solution

Choudhury et al. [173] optimized the cutting parameters for multi-responses (flank wear, surface roughness, MRR, and chip morphology) during machining of Incoloy 330 super alloy using the TOPSIS optimization method and observed it to be effective for engineering application problems. A machining operation was performed through the Taguchi L9 orthogonal array and cutting speed was found to be the dominant parameter for tool wear from ANOVA results. Pandey et al. [174] implemented the TOPSIS approach for the optimization of process parameters for multiple performance characteristics (flank wear, surface roughness, MRR, and chip morphology) during MQL machining using trihexyltetradecylphosphonium chloride ionic fluid, which was found to be effective in machining application problems. The effect of responses are studied using main effects plot, contour plot, surface plots, and regression models are also developed to find the correlations between input and output parameters. Routara et al. [175] optimized the multiple performance characteristics using the TOPSIS method during nonconventional electrical discharge machining of aluminum alloy with Cu tool in steady and rotary modes and found it to be efficient in applications.

4.4.8 AHP, ANN, Utility Theory, and Taguchi's Quality Loss Function

Das and Chattopadhyay [176] observed that the AHP (analytic hierarchy process) was reasonably good to assess the state of tool wear in the turning of C60 high carbon steel with TiN-coated carbide tool based on cutting force measurement. Cutting force and flank wear data were used to develop the AHP model. It was observed that the AHP-estimated tool wear closely matches with experimentally observed tool wear and thus can be applied safely for estimation of tool wear condition. Yuvaraju et al. [177] studied the internal turning process of hybrid nanocomposite coatings on the surface of boring bar using different tool holders (conventional, nano-SiC/GFRF with 1% SiC, nano-SiC/GFRF with 2% SiC) through the Box–Behnken DOE. The influence of input parameters on responses (amplitude of vibration and surface roughness) was analyzed through an ANN model. Furthermore, ANN was employed in simulated annealing to yield optimal parametric conditions. Sahoo et al. [178] studied machinability and optimized multi-response parameters through the utility concept. Experiments were conducted through Taguchi DOE for surface roughness parameters. A second-order mathematical model was also developed and was found to be adequate. Sahoo and Mohanty [179] conducted machining experiment through Taguchi parameter DOE considering orthogonal array for cutting force and chip reduction coefficient. The responses were optimized individually by the Taguchi approach, and for simultaneous optimization to minimize both responses, Taguchi's quality loss function technique was used and was found to be effective. Nian et al. [180] investigated machining of S45C steel bar with carbide cutting tool using the Taguchi L9 orthogonal array and optimized the multi-response parameters through normalized Taguchi loss function taking S/N ratio with different weighting factors. Confirmation tests were conducted to predict and verify the improvement of responses and was observed to be improved through proposed the methodology of multi-response optimization. It was found that feed rate and cutting speed were the significant cutting parameters that affect the multiple performance characteristics such as tool life, cutting force, and surface finish. Abhishek et al. [181] conducted machining of CFRP composites using HSS cutting tool by varying process parameters such as spindle speed, feed rate, depth of cut, and fiber orientation angle and measured MRR, surface roughness, and cutting force. Nonlinear regression models have also been developed for correlation with process parameters. Furthermore, using the TLBO algorithm, process parameters were optimized to achieve satisfactory responses. TLBO is found to be more effective compared to GA during machining of CFRP composites. Banik et al. [182] optimized multi-response parameters during EDM of titanium alloys through the application of satisfaction function and distance-based approach. Individual satisfaction of output responses was obtained through satisfaction function and then converted into a single index. By applying distance measure and distance function through the Taguchi approach, the optimal parametric combination was obtained.

4.4.9 Hybrid Optimization Techniques

Kumar et al. [183] investigated drilling of hybrid graphene nanocomposites through RSM experimental array. Multi-responses (surface roughness, mean roughness depth, and circularity error) were optimized through an integrated approach of combined compromise solution and principal component analysis (CoCoSo-PCA). The optimal parameters was found to be drill speed of 2400 rpm, feed of 80 mm/min, and 1 wt% of graphene oxide. Feed rate had a major impact on surface roughness and circularity error. An ANOVA study was carried out to find the adequacy of the hybrid model. Manikandan et al. [184] developed a hybrid gray-ANN model during EDM of hard materials. For DOE, the Taguchi methodology was used and the significance of EDM parameters on responses was studied by ANOVA. The GRG was obtained by GRA to determine multi-response optimization model. By using a gray relational coefficient as input, an ANN model was developed to predict the GRG. A comparative analysis between experimental and predicted values was carried out. Mausam et al. [185] investigated processing and EDM of multiphase CNT-reinforced nanocomposites and optimized the process parameters (peak current, gap voltage, pulse-on time, and duty cycle) using the Taguchi-fuzzy logic approach for multiple performance characteristics to minimize tool wear and maximize MRR for process productivity and validated it. Kharwar and Verma [186] described the influence of process parameters on responses (surface roughness, thrust force, and torque) during drilling of multiwall carbon nanotube/epoxy nanocomposites and optimized multi-responses by GRA and GRA-PCA hybrid approach. The prepared sample and drilling setup of MWCNT/epoxy polymer nanocomposites are shown in Figure 4.33 [186]. The flow chart of the proposed hybrid GRA-PCA optimization technique is shown in Figure 4.34 [186]. The optimal parametric combination was found to be reinforcement of 1 wt%, spindle speed of 1000 rpm, feed rate of 50 mm/min, and a TiAlN drill bit material. Feed was found to have the principal effect on machining responses. GRA-PCA hybrid approach was more effective compared to traditional GRA approach because of less average error (5.03% for GRA-PCA and 8.02% for

FIGURE 4.33 Drilling setup of MWCNT/epoxy polymer nanocomposites. (From Kharwar, P.K. and Verma, R.K., *Measurement*, 158, 107701, 2020.)

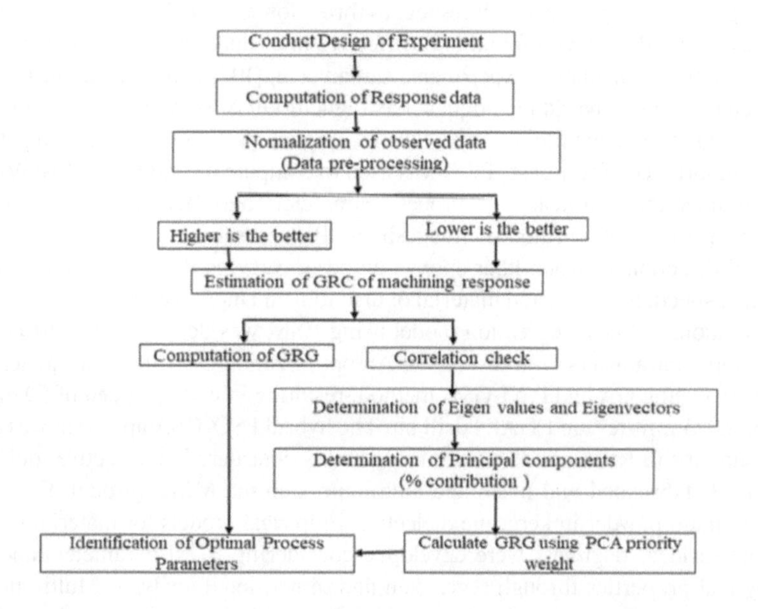

FIGURE 4.34 Flowchart for the proposed optimization approach. (From Kharwar, P.K. and Verma, R.K., *Measurement*, 158, 107701, 2020.)

GRA) and improvement in assessment values. Normal probability plot along with fitted value histogram of GRA-PCA module indicated satisfactory performance of the hybrid approach. Thus the hybrid GRA-PCA module is seen to be precise and accurate compared to traditional GRA.

Singh et al. [187] optimized multi-responses during ultrasonic machining of WC-Co composite by gray-based fuzzy logic integrated with the Taguchi approach. The experiment has been conducted using the Taguchi L36 orthogonal array design to study multi-responses such as MRR and tool wear. Grit size and power rating were found to be significant parameters for multi-responses. Gray relation fuzzy grade was computed by converting all responses into single performance through gray-based fuzzy logic. Rajesh et al. [188] studied machining of red-mud-based MMCs and optimized multi-responses (surface roughness, vibration, power consumption) simultaneously by Taguchi-based gray analysis integrated with PCA. The experiment was conducted based on the Taguchi L9 orthogonal array design. The proposed methodology of GRA-PCA is useful and effective to yield optimal parameters in CNC machining processes. Neeli et al. [189] adopted GRA and desirability function analysis (DFA) approaches for multi-response (surface roughness and delamination factor) optimization during machining (milling operation) of GFRP composites through Taguchi experimental design by L27 orthogonal array. ANOVA has been studied to find the significance of process parameters on responses. Verma et al. [190] investigated machining of GFRP composites and optimized multiple performance responses using PCA coupled with fuzzy logic and finally by Taguchi technique; quality improvement was observed through the proposed methodology. The methodology is fruitful as the Taguchi method has the capability to solve single response optimization problems. Routar et al. [191] optimized multi-response drilling parameters for GFRP composites using a gray-fuzzy approach with respect to thrust force and delamination factor. The proposed methodology was found to be effective in evaluating optimal parameters as evident from a confirmation experiment. Zaman et al. [192] optimized multi-response parameters through hybrid techniques like Taguchi-GRA-PCA during machining. In GRA, GRG is calculated to optimize through Taguchi S/N ratio. In order to get individual importance of response, PCA was used to compute overall GRG. ANOVA was done to judge the significance of process parameters on GRG and was finally validated by a confirmation run. Shunmugesh and Panneerselvam [193] conducted drilling of bi-directional carbon fiber-epoxy composite varying drilling parameters such as cutting speed, feed rate, and material of drill tool and measured thrust force, torque, and vibration. A linear regression model using RSM was developed for the correlation of input parameters with responses. An optimization of drilling parameters was carried out using GA and PSO-GSA method, resulting in cutting speed of 50 m/min, feed of 0.025 mm/rev, and TiAlN drill bit. The hybrid PSO-GSA approach was found to be superior to RSM and GA techniques and is considered an effective tool. Talla et al. [194] fabricated and machined aluminum/alumina MMC through EDM adding aluminum powder in kerosene dielectric. Empirical models for material removal rate and surface roughness were developed considering EDM parameters and thermophysical properties through regression and dimensional analysis. Multi-objective optimization of parameters to obtain minimum surface roughness and maximum MRR was performed through a hybrid approach such as Gray-PCA. The optimal

EDM parameters were powder concentration of 4 g/L, peak current of 3 A, pulse-on time of 150 μs and, duty cycle of 85%. Das et al. [195] fabricated Al-4.5% Cu/5TiC MMC by flux-assisted synthesis (FAS) and studied the effect of milling parameters (cutting speed, feed rate, and depth of cut) on responses (cutting force and COM) through ANOVA. Multi-objective optimization of parameters was performed through Taguchi-fuzzy logic. Regression models were developed for responses and verified. There was good agreement between predicted and experimental values, as the error for both outputs was below 3%. Tanvir et al. [196] conducted machining of stainless steel and studied the influence of machining parameters (cutting speed, feed rate, and depth of cut) on responses (surface roughness, cutting force, power, peak tool temperature, MRR, and heat rate). Multi-response optimization of cutting parameters was performed using hybrid whale optimization algorithm (WOA)-GRA, and an optimal condition was obtained considering unit cost and quality of production. The proposed hybrid optimization methodology was found to be suitable, having different weightage factors for each response.

Many researchers in the recent past have performed conventional and nonconventional machining experiments of various materials and optimized the process parameters for multi-responses by various multi-criteria decision-making approaches verified by confirmation runs. Techno-economical aspects during machining operation have also been investigated for sustainability. The proposed methodologies adopted have shown effectiveness and improvement. These machining experiments were performed through various DOE techniques, which are found to be popular nowadays. Furthermore, mathematical/statistical models were also developed for responses and adequacy checked through ANOVA, F-test, coefficient of correlation values (R^2), and also by normal probability plot and surface plot [196–222].

4.5 INFERENCES

The aim of this chapter is to highlight the recent progress and significant findings in the area of intelligent machining, design of experiments, and optimization techniques with emphasis on their application in the machining of nanocomposites. The computational techniques, various experimental design approaches, single- and multi-response optimization techniques, selection of appropriate parameters, statistical techniques to get significance and contribution of each parameter, and optimization of single and multiple responses were discussed thoroughly. It is clear that very little research has been undertaken for intelligent machining of nanocomposite materials, paving the need for studies related to the development of good low-cost sensors, effective computing tools, formulation of optimization algorithms for different machining areas of nanocomposites, and focusing on internet-based machining research for intelligent systems. Digital manufacturing, with collection of data through sensors, will definitely enhance productivity in machining nanocomposites. Development of real-time simulation models through machine learning and effective computational models can better illustrate machining processes of nanocomposites.

Taguchi, RSM, fuzzy-neural network hybrid technique for prediction model and optimization through soft computing approach like GA and PSO have been observed to be popular and principally preferred in conventional and nonconventional machining. However certain issues need to be addressed quickly so that they can be made suitable for industrial applications for economical and efficient acquisition of data and noisy data filtration.

Many investigations have been carried out for machining of MMCs. The challenging task is mainly because of nonuniform distribution of reinforcement in the matrix and its nonhomogeneous and anisotropic nature. Extensive research has also been undertaken for machining of fiber-reinforced composites but limited research is available for machining of natural composites and nanocomposites and their surface integrity aspects. This chapter will definitely be helpful for researchers interested in the different aspects of machining. Machining is still a challenging task and can be made economical by utilizing appropriate machining parameters, geometrical parameters, cutting tool materials, and environmental parameters.

REFERENCES

[1] S. Deb, U.S. Dixit, Intelligent machining: computational methods and optimization, in: *Machining*, 329–358. Springer, London, 2008. https://doi.org/10.1007/978-1-84800-213-5_12.

[2] M. Imad, C. Hopkins, A. Hosseini, N.Z. Yussefian, H.A. Kishawy, Intelligent machining: a review of trends, achievements and current progress, *International Journal of Computer Integrated Manufacturing*, 2021. https://doi.org/10.1080/0951192X.2021.1891573.

[3] M. Chandrasekharan, M. Muralidhar, C.M. Krishna, U.S. Dixit, Application of soft computing techniques in machining performance prediction and optimization: a literature review, *International Journal of Advanced Manufacturing Technology*, 46, 445–464, 2010.

[4] R.F. Zinati, M.R. Razfar, An investigation of the machinability of PA 6/nano-CaCO3 composite, *The International Journal of Advanced Manufacturing Technology*, 68, 2489–2497, 2013.

[5] R. Teti, T. Segreto, A. Caggiano, L. Nele, Smart multi-sensor monitoring in drilling of CFRP/CFRP composite material stacks for aerospace assembly applications, *Applied Sciences*, 10(3), 758–773, 2020.

[6] P. Asadi, M. K. B. Givi, A. Rastgoo, M. Akbari, V. Zakeri, S. Rasouli, Predicting the grain size and hardness of AZ91/SiC nanocomposite by artificial neural networks, The *International Journal of Advanced Manufacturing Technology*, 63, 1095–1107, 2012.

[7] S. Ray, A.K. Rout, A.K. Sahoo, Tribomechanical performance of glass–epoxy hybrid composites filled with marble powder with the Taguchi design and artificial neural network, *Composites: Mechanics, Computations, Applications: An International Journal*, 10(1), 17–38, 2019.

[8] E. Bagci, B. Isik, Investigation of surface roughness in turning unidirectional GFRP composites by using RS methodology and ANN, *International Journal of Advanced Manufacturing Technology*, 31, 10–17, 2006.

[9] D. Abdul Budan, Machinability study on FRP composites-A neural network analysis, *Indian Journal of Engineering & Materials Sciences*, 11, 193–200, 2004.

[10] A.C. Basheer, U.A. Dabade, S.S. Joshi, V.V. Bhanuprasad, V.M. Gadre, Modeling of surface roughness in precision machining of metal matrix composites using ANN, *Journal of Materials Processing Technology*, 197(1–3), 439–444, 2008.

[11] N. Muthukrishnan, J. Paulo Davim, Optimization of machining parameters of Al/SiC-MMC with ANOVA and ANN analysis, *Journal of Materials Processing Technology*, 209(1), 225–232, 2009.

[12] A. Panda, A.K. Sahoo, I. Panigrahi, A.K. Rout, Investigating machinability in hard turning of AISI 52100 bearing steel through performance measurement: QR, ANN and GRA study, *International Journal of Automotive and Mechanical Engineering*, 15, 4935–4961.

[13] R. Kumar, A.K. Sahoo, P.C. Mishra, R.K. Das, S. Roy, ANN modeling of cutting performances in spray cooling assisted hard turning, *Materials Today: Proceedings*, 5(9), 18482–18488, 2018.

[14] A.K. Sahoo, A.K. Rout, D. Das, Response surface and artificial neural network prediction model and optimization for surface roughness in machining, *International Journal of Industrial Engineering Computations*, 6(2), 229–240, 2015.

[15] R. Kumar, A.K. Sahoo, R.K. Das, A. Panda, P.C. Mishra, Modelling of flank wear, surface roughness and cutting temperature in sustainable hard turning of AISI D2 steel, *Procedia Manufacturing*, 20, 406–413, 2018.

[16] S. Roy, R. Kumar, A.K. Sahoo, A. Panda, Cutting tool failure and surface finish analysis in pulsating MQL-assisted hard turning, *Journal of Failure Analysis and Prevention*, 20(4), 1274–1291, 2020.

[17] C. Lu, Study on prediction of surface quality in machining process, *Journal of Materials Processing Technology*, 205(1–3), 439–450, 2008.

[18] J.P. Davim, V.N. Gaitonde, S.R. Karnik, Investigations into the effect of cutting conditions on surface roughness in turning of free machining steel by ANN models, *Journal of Materials Processing Technology*, 205(1–3), 16–23, 2008.

[19] I. Escamilla, L. Torres, P. Perez, P. Zambrano, A comparison between back propagation and the maximum sensibility neural network to surface roughness prediction in machining of Titanium (Ti 6Al 4V) alloy, in: A. Gelbukh, E.F. Morales (eds) *MICAI 2008: Advances in Artificial Intelligence. MICAI 2008*, Lecture Notes in Computer Science, vol. 5317. Springer, Berlin, Heidelberg, 2008. https://doi.org/10.1007/978-3-540-88636-5_95.

[20] D. Karayel, Prediction and control of surface roughness in CNC lathe using artificial neural network. *Journal of Materials Processing Technology*, 209(7), 3125–3137, 2009.

[21] I. Asiltürk, M. Çunkaş, Modeling and prediction of surface roughness in turning operations using artificial neural network and multiple regression method, *Expert Systems with Applications*, 38(5), 5826–5832, 2011.

[22] K.A. Risbood, U.S. Dixit, A.D. Sahasrabudhe, Prediction of surface roughness and dimensional deviation by measuring cutting forces and vibrations in turning process, *Journal of Materials Processing Technology*, 132(1–3), 203–214, 2003.

[23] T. Özel, Y. Karpat, Predictive modeling of surface roughness and tool wear in hard turning using regression and neural networks, *International Journal of Machine Tools and Manufacture*, 45(4–5), 467–479, 2005.

[24] N.K. Sahu, A.B. Andhare, S. Andhale, R.R. Abraham, Prediction of surface roughness in turning of Ti-6Al-4V using cutting parameters, forces and tool vibration, *IOP Conference Series: Materials Science and Engineering*, 346, 012037, 2018.

[25] S. Basak, U.S. Dixit, J.P. Davim, Application of radial basis function neural networks in optimization of hard turning of AISI D2 cold-worked tool steel with a ceramic tool, *Proceedings of the Institution of Mechanical Engineers Part B: Journal of Engineering Manufacture*, 221(6), 987–998, 2007.

[26] T. Özel, Y. Karpat, L. Figueira, J.P. Davim, Modelling of surface finish and tool flank wear in turning of AISI D2 steel with ceramic wiper inserts, *Journal of Materials Processing Technology*, 189(13), 192–198, 2007.

[27] M.R. Phate, S.B. Toney, Modeling and prediction of WEDM performance parameters for Al/SiCp MMC using dimensional analysis and artificial neural network, *Engineering Science and Technology, an International Journal*, 22(2), 468–476, 2019.

[28] G. Ozden, F. Mata, M.O. Oteyaka, Artificial neural network modeling for prediction of cutting forces in turning unreinforced and reinforced polyamide, *Journal of Thermoplastic Composite Materials*, 34(3), 353–363, 2021.

[29] D. Karayel, Prediction and control of surface roughness in CNC lathe using artificial neural network, *Journal of Materials Processing Technology*, 209, 3125–3137, 2009.

[30] P.V. Pashaki, M. Pouya, V.A. Maleki, High-speed cryogenic machining of the carbon nanotube reinforced nanocomposites: finite element analysis and simulation, *Proceedings of IMechE Part C: Journal of Mechanical Engineering Science*, 232(11), 1927–1936, 2018.

[31] X. Teng, D. Huo, W. Chen, E. Wong, L. Zheng, I. Shyha, Finite element modelling on cutting mechanism of nano Mg/SiC metal matrix composites considering cutting edge radius, *Journal of Manufacturing Processes*, 32, 116–126, 2018.

[32] S. Ray, A.K. Rout, A.K. Sahoo, An experimental and computational study on thermal conductivity of marble particle filled epoxy composites, *Rasayan Journal of Chemistry*, 11(1), 80–87, 2018.

[33] M. El-Gallab, M. Sklad, Machining of Al/SiC particulate metal matrix composites part III: comprehensive tool wear models, *Journal of Materials Processing Technology*, 101(1–3), 10–20, 2000.

[34] Z. Zhenzhong, Y. Peng, W. Jun, H. Chuanzhen, R. Cai, Z. Hongtao, Analytical modeling of surface roughness in precision grinding of particle reinforced metal matrix composites considering nanomechanical response of material, *International Journal of Mechanical Sciences*, 157–158, 243–253, 2019.

[35] J.P. Davim, C.A. Conceicao Antonio, Optimization of cutting conditions in machining of aluminium matrix composites using a numerical and experimental model, *Journal of Materials Processing Technology*, 112, 78–82, 2001.

[36] L. Tang, J. Huang, L. Xie, Finite element modeling and simulation in dry hard orthogonal cutting AISI D2 tool steel with CBN cutting tool, *International Journal of Advanced Manufacturing Technology*, 53, 1167–1181, 2011.

[37] U. Umer, M.H. Abidi, J.A. Qudeiri, H. Alkhalefah, H. Kishawy, Tool performance optimization while machining aluminium-based metal matrix composite, *Metals*, 10, 835–850, 2020.

[38] A. A. Elsadek, A. M. Gaafer, M. A. Lashin, Prediction of roughness of and tool wear in turning of metal matrix nanocomposites, *Journal of Engineering and Applied Sciences*, 64(5), 387–408, 2017.

[39] S. Sharma, S. Tamang, D. Devarasiddappa, M. Chandrasekharan, Fuzzy logic modeling and multiple performance optimization in turning GFRP composites using desirability function analysis, *Procedia Materials Science*, 6, 1805–1814, 2014.

[40] Y. Jiao, S. Lei, Z.J. Pei, E.S. Lee, Fuzzy adaptive networks in machining process modeling: surface roughness prediction for turning operations, *International Journal of Machine Tools and Manufacture*, 44(15), 1643–1651, 2004.

[41] N.R. Abburi, U.S. Dixit, A knowledge-based system for the prediction of surface roughness in turning process, *Robotics and Computer-Integrated Manufacturing*, 22(4), 363–372, 2006.

[42] S. Ramesh, L. Karunamoorthy, K. Palanikumar, Fuzzy modeling and analysis of machining parameters in machining titanium alloy, *Materials and Manufacturing Processes*, 23(4), 439–447, 2008.

[43] E.D. Kirby, J.C. Chen, Development of a fuzzy-nets-based surface roughness prediction system in turning operations, *Computers & Industrial Engineering*, 53(1), 30–42, 2007.

[44] B.C. Routara, A.K. Sahoo, A.K. Parida, P.C. Padhi, Response surface methodology and genetic algorithm used to optimize the cutting condition for surface roughness parameters in CNC turning, *Procedia Engineering*, 38, 1893–1904, 2012.

[45] S.K. Tamang, M. Chandrasekharan, A.K. Sahoo, Sustainable machining: an experimental investigation and optimization of machining Inconel 825 with dry and MQL approach, *Journal of the Brazilian Society of Mechanical Sciences and Engineering*, 40(8), 1–18, 2018.

[46] M.R.S Yazdi, M.R. Razfar, M. Asadnia, Modelling of the thrust force of the drilling operation on PA6–nanoclay nanocomposites using particle swarm optimization, *Proceedings of IMechE Part B: Journal of Engineering Manufacture*, 225(10), 1757–1771, 2011.

[47] M. Moghri, M. Madic, M. Omidi, M. Farahnakian, Surface roughness optimization of Polyamide-6/Nanoclay nanocomposites using artificial neural network: genetic algorithm approach, *The Scientific World Journal*, Article ID 485205, 2014, https://doi.org/10.1155/2014/485205.

[48] A. Fathy, A. A. Megahed, Prediction of abrasive wear rate of in situ Cu–Al2O3 nanocomposite using artificial neural networks, *The International Journal of Advanced Manufacturing Technology*, 62, 953–963, 2012.

[49] M.M. Nasr, S. Anwar, A.M.A. Samhan, M. Ghaleb, A. Dabwan, Milling of Graphene reinforced Ti6Al4V nanocomposites: an artificial intelligence based industry 4.0 approach, *Materials (Basel)*, 13(24), 5707, 2020.

[50] A. Krishnamoorthy, S. Rajendra Boopathy, K. Palanikumar, J. Paulo Davim, Application of grey fuzzy logic for the optimization of drilling parameters for CFRP composites with multiple performance characteristics, *Measurement*, 45(5), 1286–1296, 2012.

[51] S. Li, Y. Su, X. Zhu, H. Jin, Q. Ouyang, D. Zhang, Enhanced mechanical behavior and fabrication of silicon carbide particles covered by in-situ carbon nanotube reinforced 6061 aluminum matrix composites, *Materials & Design*, 107, 130–138, 2016.

[52] R. Karthikeyan, R. Adalarasan, B.C. Pai, Optimization of machining characteristics for Al/SiCp composites using ANN/GA, *Journal of Materials Science and Technology*, 18(1), 47–50, 2002.

[53] R. Kumar, A. Pandey, A. Panda, R. Mallick, A.K. Sahoo, Grey-fuzzy hybrid optimization and cascade neural network modelling in hard turning of AISI D2 steel, *International Journal of Integrated Engineering*, 13(4), 189–207, 2021.

[54] N.C. Tsourveloudis, Predictive modeling of the Ti6Al4V alloy surface roughness, *Journal of Intelligent & Robotic Systems*, 60(3–4), 513–530, 2010.

[55] S. Swain, I. Panigrahi, A.K. Sahoo, A. Panda, Adaptive tool condition monitoring system: a brief review, *Materials Today: Proceedings*, 23, 474–478, 2020.

[56] S. Swain, I. Panigrahi, A.K. Sahoo, A. Panda, R. Kumar, Effect of tool vibration on flank wear and surface roughness during high-speed machining of 1040 steel, *Journal of Failure Analysis and Prevention*, 20, 976–994, 2020.

[57] A. Panda, A.K. Sahoo, I. Panigrahi, A.K. Rout, Prediction models for on-line cutting tool and machined surface condition monitoring during hard turning considering vibration signal, *Mechanics & Industry*, 21(5), 520–535, 2020.

[58] A. Panda, A.K. Sahoo, I. Panigrahi, R. Kumar, Tool condition monitoring during hard turning of AISI 52100 Steel: a case study, *Materials Today: Proceedings*, 5(9), 18585–18592, 2018.

[59] S. Swain, S. Chand, I. Panigrahi, A.K. Sahoo, Cutting tool vibration analysis for better surface finish during dry turning of mild steel, *Materials Today: Proceedings*, 5(11), 24605–24611, 2018.

[60] V. Upadhyay, P.K. Jain, N.K. Mehta, In-process prediction of surface roughness in turning of Ti–6Al–4V alloy using cutting parameters and vibration signals, *Measurement*, 46(1), 154–160, 2013.

[61] B.Y. Lee, Y.S. Tarng, Surface roughness inspection by computer vision in turning operations. *International Journal of Machine Tools and Manufacture*, 41(9), 1251–1263, 2001.

[62] C.K.H. Dharan, M.S. Won, Machining parameters for an intelligent machining system for composite laminates, *International Journal of Machine Tools and Manufacture*, 40(3), 415–426, 2000.

[63] V.P. Bhemuni and S.R. Chalamalasetti, A Review on hard turning by using design of experiments, *Journal for Manufacturing Science and Production*, 13(3), 209–219, 2013.

[64] R. Myers, D.C. Montgomery, *Response Surface Methodology*, 2nd ed. Wiley, New York, 2002.

[65] M. Nalbant, H. Gokkaya, G. Sur, Application of Taguchi method in the optimization of cutting parameters for surface roughness in turning, *Materials and Design*, 28, 1379–1385, 2007.

[66] P.J. Ross, *Taguchi Technique for Quality Engineering*. McGraw-Hill, New York, 1988.

[67] J.Z. Zhang, J.C. Chen, E.D. Kirby, Surface roughness optimization in an end-milling operation using the Taguchi design method, *Journal of Materials Processing Technology*, 184, 233–239, 2007.

[68] D.A. Fadare, W.F. Sales, E.O. Ezugwu, J. Bonney, A.O. Oni, Effects of cutting parameters on surface roughness during high-speed turning of Ti-6AI-4V alloy, *Journal of Applied Sciences Research*, 5(7), 757–764, 2009.

[69] D. Priyadarshi, R.K. Sharma, Optimization for turning of Al-6061-SiC-Gr hybrid nanocomposites using response surface methodologies, *Materials and Manufacturing Processes*, 31(10), 1342–1350, 2016.

[70] S. Gopalakannan, T. Senthilvelan, A parametric study of electrical discharge machining process parameters on machining of cast Al/B_4C metal matrix nanocomposites, *Proceedings of IMechE Part B: Journal of Engineering Manufacture*, 227(7), 993–1004, 2013.

[71] M. Hourmand, S. Farahany, A. A. D. Sarhan, M.Y. Noordin, Investigating the electrical discharge machining (EDM) parameter effects on Al-Mg2Si metal matrix composite (MMC) for high material removal rate (MRR) and less EWR–RSM approach, *The International Journal of Advanced Manufacturing Technology*, 77, 831–838, 2015.

[72] Y.Gong, Y-J. Baik, C.P. Li, C. Byon, J.M. Park, Experimental and modeling investigation on machined surfaces of HDPE-MWCNT polymer nanocomposite, *The International Journal of Advanced Manufacturing Technology*, 88, 879–885, 2017.

[73] S. Gopalakannan, T. Senthilvelan, Application of response surface method on machining of Al–SiC nano-composites, *Measurement*, 46(8), 2705–2715, 2013.

[74] S. Mohanty, B.C. Routara, B.K. Nanda, D.K. Das, A.K. Sahoo, Study of machining characteristics of Al-SiCp 12% composite in nano powder mixed dielectric electrical discharge machining using RSM, *Materials Today: Proceedings*, 5(11), 25581–25590, 2018.

[75] M.Y. Noordin, V.C. Venkatesh, S. Sharif, S. Elting, A. Abdullah, Application of response surface methodology in describing the performance of coated carbide tools when turning AISI 1045 steel, *Journal of Materials Processing Technology*, 145, 46–58, 2004.

[76] Y.T. Ic, E.S. Güler, B. Sezer, B.S. Taş, H.S. Şahin, Multi-objective optimization of turning parameters for SiC- or Al2O3-reinforced aluminum matrix composites, *Process Integration and Optimization for Sustainability*, 2021, https://doi.org/10.1007/s41660-021-00169-4.

[77] P.V.S. Suresh, P. Venkateswara Rao, S.G. Deshmukh, A genetic algorithmic approach for optimization of surface roughness prediction model, *International Journal of Machine Tools & Manufacture*, 42, 675–680, 2002.

[78] I.A. Choudhury, M.A. El-Baradie, Tool-life prediction model by design of experiments for turning high strength steel (290 BHN), *Journal of Materials Processing Technology*, 77, 319–326, 1998.

[79] M. A. Dabnun, M.S.J. Hashmi, M.A. El-Baradie, Surface roughness prediction model by design of experiments for turning machinable glass–ceramic (Macor), *Journal of Materials Processing Technology*, 164–165, 1289–1293, 2005.
[80] I. Puertas Arbizu, C.J. Luis Pérez, Surface roughness prediction by factorial design of experiments in turning processes, *Journal of Materials Processing Technology*, 143–144, 390–396, 2003.
[81] C-X (Jack) Feng, An experimental study of the impact of turning parameters on surface roughness, Proceedings of the 2001 Industrial Engineering Research Conference, Institute of Industrial Engineers, Paper No. 2036, 1–10, 2001.
[82] M. Nalbant, H. Gokkaya, I. Toktas, Comparison of regression and artificial neural network models for surface roughness prediction with the cutting parameters in CNC turning, *Modelling and Simulation in Engineering*, 2007, 1–14, 2007, doi:10.1155/2007/92717.
[83] T. Özel, A. Esteves Correia, J. P. Davim, Neural network process modelling for turning of steel parts using conventional and wiper Inserts, *International Journal of Materials and Product Technology*, 35(1/2), 246–258, 2009.
[84] P. Thangavel, V. Selladurai, An experimental investigation on the effect of turning parameters on surface roughness, *International Journal of Manufacturing Research*, 3(3), 285–300, 2008.
[85] H. Singh, P. Kumar, Mathematical models of tool life &surface roughness for turning operation through response surface methodology, *Journal of Scientific & Industrial Research*, 66, 220–226, 2007.
[86] A.M.A. Al-Ahmari, Predictive machinability models for a selected hard material in turning operations, *Journal of Materials Processing Technology*, 190, 305–311, 2007.
[87] V.N. Gaitonde, S.R. Karnik, L. Figueira, J.P. Davim, Machinability investigations in hard turning of AISI D2 cold work tool steel with conventional and wiper ceramic inserts, *International Journal of Refractory Metals & Hard Materials*, 27(2), 754–763, 2009.
[88] D.I. Lalwani, N.K. Mehta, P.K. Jain, Experimental investigations of cutting parameters influence on cutting forces and surface roughness in finish hard turning of MDN250 steel, *Journal of Materials Processing Technology*, 206(1–3), 167–179, 2008.
[89] R.S. Rana, R. Purohit, P.M. Mishra, P. Sahu, S. Dwivedi, Optimization of mechanical properties of AA 5083 nano SiC composites using design of experiment technique, *Materials Today: Proceedings*, 4, 3882–3890, 2017.
[90] A.K. Sahoo, P.C. Mishra, A response surface methodology and desirability approach for predictive modeling and optimization of cutting temperature in machining hardened steel, *International Journal of Industrial Engineering Computations*, 5(3), 407–416, 2014.
[91] A.K. Sahoo, K. Orra, B.C. Routra, Application of response surface methodology on investigating flank wear in machining hardened steel using PVD TiN coated mixed ceramic insert, *International Journal of Industrial Engineering Computations*, 4(4), 469–478, 2013.
[92] M. Czampa, I. Biro, T. Szalay, Effects of different cutting conditions on the surface roughness parameters of iron-copper-carbon powder metallurgy composites, *International Journal of Machining and Machinability of Materials*, 19(5), 440–456, 2017.
[93] A. Sharma, M.P. Garg, K.K. Goyal, A. Kumar, Investigation of wire electrical discharge machining of ZrSiO4p/Al 6063 MMC, *International Journal of Machining and Machinability of Materials*, 18(4), 392–411, 2016.
[94] H. Kumar, R. Kumar, A. Manna, Effects of electrode configuration on MRR and EWR during electric discharge machining of Al/10wt%SiCp-MMC, *International Journal of Machining and Machinability of Materials*, 18(1/2), 54–76, 2016.

[95] V. Kumar, M. Kharub, A. Sinha, Modeling and Optimization of Turning Parameters during Machining of AA6061 composite using RSM Box-Behnken Design, IOP Conference Series: Materials Science and Engineering, 1057, 012058, 2021.

[96] A. Hasçalık, U. Çaydaş, Optimization of turning parameters for surface roughness and tool life based on the Taguchi method, *The International Journal of Advanced Manufacturing Technology*, 38(9–10), 896–903, 2008.

[97] M.V. Ramana, G.K.M. Rao, D.H. Rao, Optimization and effect of process parameters on tool wear in turning of titanium alloy under different machining conditions, International Journal of Materials, *Mechanics and Manufacturing*, 2(4), 272–277, 2014.

[98] A.K. Sahoo, Application of Taguchi and regression analysis on surface roughness in machining hardened AISI D2 steel, *International Journal of Industrial Engineering Computations*, 5(2), 295–304, 2014.

[99] R. Shetty, R.B. Pai, S.S. Rao, R. Nayak, Taguchi's technique in machining of metal matrix composites, *Journal of the Brazilian Society of Mechanical Sciences and Engineering*, 31(1), 2009, https://doi.org/10.1590/S1678-58782009000100003.

[100] J.P. Davim, Design of optimization of cutting parameters for turning metal matrix composites based on the orthogonal arrays, *Journal of Materials Processing Technology*, 132, 340–344, 2003.

[101] K. Palanikumar, Application of Taguchi and response surface methodologies for surface roughness in machining glass fiber reinforced plastics by PCD tooling, *International Journal of Advanced Manufacturing Technology*, 36, 19–27, 2008, https://doi.org/10.1007/s00170-006-0811-0.

[102] J.P. Davim, P. Reis, Machinability study on composite (polyetheretherketone reinforced with 30% glass fifibre–PEEK GF 30) using polycrystalline diamond (PCD) and cemented carbide (K20) tools, *International Journal of Advanced Manufacturing Technology*, 23, 412–418, 2004.

[103] K. Palanikumar, Cutting parameters optimization for surface roughness in machining of GFRP composites using Taguchi's method, *Journal of Reinforced Plastics and Composites*, 25(16), 1739–1751, 2006.

[104] S.A.Hussain, V. Pandurangadu, K.Palanikumar, Surface roughness analysis in machining of GFRP composites by carbide tool (K20), *European Journal of Scientific Research*, 41(1), 84–98, 2010.

[105] A. Satapathy, A. Patnaik, Analysis of dry sliding wear behavior of red mud filled polyester composites using the Taguchi method, *Journal of Reinforced Plastics and Composites*, 29(19), 2883–2897, 2010.

[106] A. Patnaik, A. Satapathy, S.S. Mahapatra, R.R. Dash, Parametric optimization erosion wear of polyester-GF-alumina hybrid composites using the Taguchi method, *Journal of Reinforced Plastics and Composites*, 27(10), 1039–1058, 2008.

[107] S.R. Dakarapu, R. Nallu, Process parameters optimization for producing AA6061/TiB_2composites by friction stir processing, *Journal of Mechanical Engineering*, 67(1), 101–118, 2017.

[108] J. Paulo Davim, A note on the determination of optimal cutting conditions for surface finish obtained in turning using design of experiment, *Journal of Materials Processing Technology*, 116, 305–308, 2001.

[109] H. Singh, P. Kumar, Optimizing feed force for turned parts through the Taguchi Technique, *Sadhana*, 31(6), 671–681, 2006.

[110] A.I Gusri, C.H Che Hassan, A.G Jaharah, B. Yanuar, A. Yasir, A. Nagi, Application of Taguchi method in optimizing turning parameters of titanium alloy, Seminar on Engg. Mathematics, Engineering Mathematics Group, 57–63, 2008.

[111] U.A. Dabade, S.S. Joshi, Analysis of chip formation mechanism in machining of Al/SiCp metal matrix composites, *Journal of Materials Processing Technology*, 209(10), 4704–4710, 2009.

[112] A.K. Sahoo, S. Pradhan, Modeling and optimization of Al/SiCp MMC machining using Taguchi approach, *Measurement*, 46(9), 3064–3072, 2013.
[113] A.K. Sahoo, S. Pradhan, A.K. Rout, Development and machinability assessment in turning Al/SiCp-metal matrix composite with multilayer coated carbide insert using Taguchi and statistical techniques, *Archives of Civil and Mechanical Engineering*, 13, 27–35, 2013.
[114] A.K. Parida, R. Das, A.K. Sahoo, B.C. Routara, Optimization of cutting parameters for surface roughness in machining of GFRP composites with graphite/fly ash filler, *Procedia Materials Science*, 6, 1533–1538, 2014.
[115] A. Rout, A. Satapathy, S. Mantry, A.K. Sahoo, T. Mohanty, Erosion wear performance analysis of polyester-GF-granite hybrid composites using the Taguchi method, *Procedia Engineering*, 38, 1863–1882, 2012.
[116] S. Ray, A.K. Rout, A.K. Sahoo, Development and characterization of glass/polyester composites filled with industrial wastes using statistical techniques, *Indian Journal of Engineering and Materials Sciences*, 25(2), 169–182, 2018.
[117] S. Ray, A.K. Rout, A.K. Sahoo, A study on erosion performance analysis of glass-epoxy composites filled with marble waste using artificial neural network, *UPB Scientific Bulletin, Series B: Chemistry and Materials Science*, 80(4), 181–196, 2018.
[118] D. Das, R.K. Thakur, A.K. Chaubey, A.K. Sahoo, Optimization of machining parameters and development of surface roughness models during turning Al-based metal matrix composite, *Materials Today: Proceedings*, 5(2), 4431–4437, 2018.
[119] P.K. Tiwari, R. Kumar, A.K. Sahoo, A. Panda, D. Das, S. Roy, Performance evaluation of coated cermet insert in hard turning, Materials Today: Proceedings, 26, 1941–1947, 2020.
[120] R.R. Mishra, R. Kumar, A.K. Sahoo, A. Panda, Machinability behaviour of biocompatible Ti-6Al-4V ELI titanium alloy under flood cooling environment, *Materials Today: Proceedings*, 23, 536–540, 2020.
[121] A.K. Sahoo, B. Sahoo, Surface roughness model and parametric optimization in finish turning using coated carbide insert: response surface methodology and Taguchi approach, *International Journal of Industrial Engineering Computations*, 2(4), 819–830, 2011.
[122] A. Manna, B. Bhattacharyya, Investigation for optimal parametric combination for achieving better surface finish during turning of Al /SiC-MMC, *International Journal of Advanced Manufacturing Technology*, 23, 658–665, 2004.
[123] S. Ramabalan, H.B. Michael Rajan, I. Dinaharan, S.J. Vijay, Experimental investigation of MRR on in situ formed AA7075/TiB2 cast composites machining by wire EDM, *International Journal of Machining and Machinability of Materials*, 17(3/4), 295–318, 2015.
[124] A. Patnaik, A. Satapathy, S.S. Mahapatra, R.R. Dash, Tribo-performance of polyester hybrid composites: damage assessment and parameter optimization using Taguchi design, *Materials & Design*, 30(1), 57–67, 2009.
[125] S.S. Mahapatra, A. Patnaik, A. Satapathy, Taguchi method applied to parametric appraisal of erosion behavior of GF-reinforced polyester composites, *Wear*, 265, 214–222, 2008.
[126] A. Patnaik, A. Satapathy, S.S. Mahapatra, R.R. Dash, Modeling and prediction of erosion response of glass reinforced polyester-flyash composites, *Journal of Reinforced Plastics and Composites*, 28(5), 513–536, 2009.
[127] A. Patnaik, A. Satapathy, S.S. Mahapatra, R.R. Dash, Implementation of Taguchi design for erosion of fiber-reinforced polyester composite systems with SiC filler, *Journal of Reinforced Plastics and Composites*, 27(10), 1093–1111, 2008.
[128] A. Patnaik, A. Satapathy, S.S. Mahapatra, R.R. Dash, A modeling approach for prediction of erosion behavior of glass fiber–polyester composites, *Journal of Polymer Research*, 15, 147–160, 2008.

[129] A. Patnaik, A. Satapathy, S.S. Mahapatra, R.R. Dash, A. Taguchi approach for investigation of erosion of glass fiber – polyester composites, *Journal of Reinforced Plastics and Composites*, 27(8), 871–888, 2008.

[130] A. Patnaik, A. Satapathy, S.S. Mahapatra, Modified erosion wear characteristics of glass-polyester composites by Silicon Carbide filling: a parametric study using Taguchi technique, *Journal of Engineering Materials and Technology*, 131(3), 031011, 2009.

[131] R.A. Ramnath, P.R. Thyla, N.M. Kumar, S. Aravind, Optimization of machining parameters of composites using multi-attribute decision-making techniques: a review, *Journal of Reinforced Plastics and Composites*, 37(2), 77–89, 2018.

[132] A. Roushan, A. Bandyopadhyay, S. Banerjee, Multiple performance characteristics optimisation in side and face milling of glass fibre reinforced polyester composite at different weightage of performances by grey relational analysis, *International Journal of Machining and Machinability of Materials*, 19(1), 41–56, 2017.

[133] B.M.C Rajan, A.Senthil Kumar, T.Sornakumar, P. Senthamaraikannan, M.R.Sanjay, Multi response optimization of fabrication parameters of carbon fiber-reinforced aluminium laminates (CARAL): by Taguchi method and Grey relational analysis, *Polymer Composites*, 40(S2), E1041–E1048, 2019.

[134] P.A. Sylajakumari, R. Ramakrishnasamy, G. Palaniappan, Taguchi grey relational analysis for multi-response optimization of wear in Co-Continuous composite, *Materials*, 11, 1743–1759, 2018.

[135] P.K. Kopparthi, V.R. Kundavarapu, V.R. Kaki, B.R. Pathakokila, Modeling and multi response optimization of mechanical properties for E-glass/polyester composite using Taguchi-grey relational analysis, *Proceedings of the IMechE, Part E: Journal of Process Mechanical Engineering*, 235(2), 342–350, 2021.

[136] H. Siddhi Jailani, A. Rajadurai, B. Mohan, A. Senthil Kumar, T. Sornakumar, Multi-response optimisation of sintering parameters of Al–Si alloy/fly ash composite using Taguchi method and grey relational analysis, *International Journal of Advanced Manufacturing Technology*, 45, 362–369, 2009.

[137] Palanikumar, Experimental investigation and optimisation in drilling of GFRP composites, *Measurement*, 44(10), 2138–2148, 2011.

[138] B.C. Routara, B.K. Nanda, A.K. Sahoo, D.N. Thatoi, B.B. Nayak, Optimisation of multiple performance characteristics in abrasive jet machining using grey relational analysis, *International Journal of Manufacturing Technology and Management*, 24(1–4), 4–22, 2011.

[139] P.C. Mishra, D.K. Das, S.K. Sahu, Comparative performance in hard turning of AISI 1015 steel with carbide insert using orthogonal array design and grey relational analysis under spray impingement cooling and dry environment: a case study, *International Journal of Manufacturing Materials and Mechanical Engineering*, 4(3), 1–32, 2014.

[140] S.C. Katamreddy, N. Punnath, N. Radhika, Multi-response optimisation of machining parameters in electrical discharge machining of Al LM25/AlB2 functionally graded composite using grey relation analysis, *International Journal of Machining and Machinability of Materials*, 20(3), 193–213, 2018.

[141] S. Datta, A. Bandyopadhyay, P.K. Pal, Modeling and optimization of features of bead geometry including percentage dilution in submerged arc welding using mixture of fresh flux and fused slag, *International Journal of Advanced Manufacturing Technology*, 36, 1080–1090, 2008.

[142] A. Noorul Haq, P. Marimuthu, R. Jeyapaul, Multi response optimization of machining parameters of drilling Al/SiC metal matrix composite using grey relational analysis in the Taguchi method, *International Journal of Advanced Manufacturing Technology*, 37, 250–255, 2008.

[143] C.L. Lin, Use of the Taguchi Method and Grey Relational Analysis to optimize turning operations with multiple performance characteristics, *Materials & Manufacturing Processes*, 19(2), 209–220, 2004.

[144] C-J Tzenga, Y-H Linb, Y-K Yanga, M-C Jeng, Optimization of turning operations with multiple performance characteristics using the Taguchi method and Grey relational analysis, *Journal of Materials Processing Technology*, 209, 2753–2759, 2009.

[145] P. Mishra, D. Das, M. Ukamanal, B. Routara, A.K. Sahoo, Multi-response optimization of process parameters using Taguchi method and grey relational analysis during turning AA 7075/SiC composite in dry and spray cooling environments, *International Journal of Industrial Engineering Computations*, 6(4), 445–456, 2015.

[146] J. Kumar, R.K. Verma, A.K. Mondal, Predictive modeling and machining performance optimization during drilling of polymer nanocomposites reinforced by graphene oxide/carbon fiber, *Archive of Mechanical Engineering*, 67(2), 229–258, 2020.

[147] P.K. Swain, K.D. Mohapatra, R. Das, A.K. Sahoo, A. Panda, Experimental investigation into characterization and machining of Al + SiCp nano-composites using coated carbide tool, *Mechanics & Industry*, 21, 307, 2020.

[148] M. Ukamanal, P.C. Mishra, A.K. Sahoo, Effects of spray cooling process parameters on machining performance AISI 316 steel: a novel experimental technique, *Experimental Techniques*, 44(1), 19–36, 2020.

[149] A. Panda, A.K. Sahoo, A.K. Rout, Investigations on surface quality characteristics with multi-response parametric optimization and correlations, *Alexandria Engineering Journal*, 55(2), 1625–1633, 2016.

[150] R.Kumar, A. Modi, A. Panda, A.K. Sahoo, A. Deep, P.K. Behera, R. Tiwari, Hard turning on JIS S45C structural steel: an experimental, modelling and optimisation approach, *International Journal of Automotive and Mechanical Engineering*, 16, 7315–7340, 2019.

[151] H.C. Liao, Multi-response optimization using weighted principal component. *The International Journal of Advanced Manufacturing Technology*, 27(7–8), 720–725, 2006.

[152] G.K. Kumar, M. R. Ch, V.V.S. KesavaRao, Application of WPCA & CQL methods in the optimization of multiple responses. *Materials Today: Proceedings*, 18(1), 25–36, 2019.

[153] P.Sahoo, A. Pratap, A. Bandyopadhyay, Modeling and optimization of surface roughness and tool vibration in CNC turning of Aluminum alloy using hybrid RSM-WPCA methodology, *International Journal of Industrial Engineering Computations*, 8(3), 385–398, 20–17.

[154] S. Roy, R. Kumar, A.K. Sahoo, A. Pandey, A. Panda, Investigation on hard turning temperature under a novel pulsating MQL environment: an experimental and modelling approach, *Mechanics & Industry*, 21(6), 605, 2020.

[155] D. Das, V. Chakraborty, B.K. Nanda, B.C. Routara, Turning performance of Al 7075/SiCp MMC and multi-response optimization using WPCA and Taguchi approach, *Materials Today: Proceedings*, 5(2), 6030–6037, 2018.

[156] D.R. Pamuji, M.A. Wahid, A. Rohman, A.A. Sonief, M.A. Choiron, Optimization of multiple response using Taguchi-WPCA in ST 60 tool steel turning process with minimum quantity cooling lubrication (MQCL) method, *Aceh International Journal of Science and Technology*, 7(1), 44–55, 2018.

[157] P. Jayaraman, L.M. Kumar, Multi-characteristics optimization during turning of AISI 6061 T6 aluminium alloy using weighted principal component analysis, *International Journal of Applied Engineering Research*, 10(8), 2015.

[158] S.R. Rao, G. Padmanabhan, Multi-response optimization of electrochemical machining of Al-Si/B4C composites using RSM, *International Journal of Manufacturing Materials and Mechanical Engineering*, 3(3), 42–56, 2013.

[159] R.K. Bhusan, Multiresponse optimization of Al alloy-SiC composite machining parameters for minimum tool wear and maximum metal removal rate, *Journal of Manufacturing Science and Engineering*, 135(2), 021013, 2013.

[160] S. Khamel, N. Ouelaa, K. Bouacha, Analysis and prediction of tool wear, surface roughness and cutting forces in hard turning with CBN tool, *Journal of Mechanical Science and Technology*, 26(11), 3605–3616, 2012.

[161] A. Khajuria, R. Bedi, B. Singh, M. Akhtar, EDM machinability and parametric optimisation of 2014Al/Al_2O_3 composite by RSM, *International Journal of Machining and Machinability of Materials*, 20(6), 536–555, 2018.

[162] J.T. Horng, N.M. Liu, K.T. Chiang, Investigating the machinability evaluation of Hadfield steel in the hard turning with Al_2O_3/TiC mixed ceramic tool based on the response surface methodology, *Journal of Materials Processing Technology*, 208, 532–541, 2008.

[163] R.K. Bhushan, Optimization of cutting parameters for minimizing power consumption and maximizing tool life during machining of Al alloy SiC particle composites, *Journal of Cleaner Production*, 39, 242–254, 2013.

[164] S. Dhandapani, T. Rajmohanr, D. Vijayan, K. Palanikumar, Multi response optimisation of machining parameters in EDM of dual particle (MWCNT + B4C) reinforced sintered composites, *International Journal of Machining and Machinability of Materials*, 20(5), 425–446, 2018.

[165] N.A. Fountas, G.V. Seretis, D.E. Manolakos, C.G. Provatidis, N.M. Vaxevanidis, Multi-objective statistical analysis and optimisation in turning of aluminium matrix particulate composite using genetic algorithms, *International Journal of Machining and Machinability of Materials*, 20(3), 236–251, 2018.

[166] S. Dhandapani, T. Rajmohanr, D. Vijayan, K. Palanikumar, Multiresponse optimisation of machining parameters in EDM of dual particle (MWCNT+B4C) reinforced sintered composites, *International Journal of Machining and Machinability of Materials*, 20(5), 425–446, 2018.

[167] S. Kumar, M. Gupta, P.S. Satsangi, Multiple-response optimization of cutting forces in turning of UD-GFRP composite using Distance-Based Pareto Genetic Algorithm approach, *Engineering Science and Technology, an International Journal*, 18(4), 680–695, 2015.

[168] D. Das, V. Chakraborty, B.B. Nayak, M.P. Satpathy, C. Samal, Machining of aluminium-based metal matrix composite - a particle swarm optimisation approach, *International Journal of Machining and Machinability of Materials*, 22(1), 79–97, 2020.

[169] R.R. Mishra, R. Kumar, A. Panda, A. Pandey, A.K. Sahoo, Particle swarm optimization of multi-responses in hard turning of D2 steel, progress in computing, analytics and networking, *Advances in Intelligent Systems and Computing*, 1119, 237–244, 2020.

[170] B.O.P. Soepangkat, R. Norcahyo, M.K. Effendi, B. Pramujati, Multi-response optimization of carbon fiber reinforced polymer (CFRP) drilling using back propagation neural network-particle swarm optimization (BPNN-PSO), Engineering Science and Technology, an International Journal, 23(3), 700–713, 2020.

[171] J.H. Biswas, Jagadish, A. Ray, Experimental investigation and optimisation of ultrasonic machining parameters on zirconia composite, *International Journal of Machining and Machinability of Materials*, 21(1/2), 115–137, 2019.

[172] V.R. Pathapalli, V.R. Basam, S.K. Gudimetta, M.R. Koppula, Optimization of machining parameters using WASPAS and MOORA, *World Journal of Engineering*, 17/2, 237–246, 2020.

[173] A.R. Choudhury, R. Kumar, A.K. Sahoo, A. Panda, A. Malakar, Machinability investigation on novel Incoloy 330 super alloy using coconut oil based SiO_2 nano fluid, *International Journal of Integrated Engineering*, 12(4), 145–160, 2020.

[174] A. Pandey, R. Kumar, A.K. Sahoo, A. Paul, A. Panda, Performance analysis of tri-hexyltetradecylphosphonium chloride ionic fluid under MQL condition in hard turning, *International Journal of Automotive and Mechanical Engineering*, 17(1), 7629–7647, 2020.

[175] B.C. Routara, D. Das, M.P. Satpathy, B.K. Nanda, A.K. Sahoo, S.S. Singh, Investigation on machining characteristics of T6-Al7075 during EDM with Cu tool in steady and rotary mode, *Materials Today: Proceedings*, 26, 2143–2150, 2020.

[176] S. Das, A.B. Chattopadhyay, Application of the analytic hierarchy process for estimating the state of tool wear, *International Journal of Machine Tools & Manufacture*, 43, 1–6, 2003.
[177] B.A.G. Yuvaraju, B.K. Nanda, J. Srinivas, Optimal cutting state predictions in internal turning operation with nano-SiC/GFRE composite layered boring tools, *International Journal of Machining and Machinability of Materials*, 23(1), 1–20, 2021.
[178] A.K. Sahoo, A. Panda, B.B. Nayak, R. Kumar, R.K. Das, R.K. Nayak, Machinability model and multi-response optimisation of process parameters through regression and utility concept, *International Journal of Process Management and Benchmarking*, 11(3), 390–414, 2021.
[179] A.K. Sahoo, T. Mohanty, Optimization of multiple performance characteristics in turning using Taguchi's quality loss function: an experimental investigation, *International Journal of Industrial Engineering Computations*, 4(3), 325–336, 2013.
[180] C.Y. Nian, W.H. Yang, Y.S. Tarng, Optimization of turning operations with multiple performance characteristics, *Journal of Materials Processing Technology*, 95, 90–96, 1999.
[181] K. Abhishek, V.R. Kumar, S. Datta, S.S. Mahapatra, Parametric appraisal and optimization in machining of CFRP composites by using TLBO (teaching–learning based optimization algorithm), *Journal of Intelligent Manufacturing*, 28(8), 1769–1785, 2017.
[182] D. Banik, Rahul, G. Kar, B. Debnath, B.C. Routara, A.K. Sahoo, D. Kochar, Machining performance optimization during electro discharge machining on Titanium (Grade 4): application of satisfaction function and distance-based approach, in K. Shanker, R. Shankar, R. Sindhwani (eds) *Advances in Industrial and Production Engineering*, Lecture Notes in Mechanical Engineering, Springer, Singapore, pp. 535–542, 2019. https://doi.org/10.1007/978-981-13-6412-9_52.
[183] J. Kumar, R.K. Verma, A novel methodology of Combined Compromise Solution and Principal Component Analysis (CoCoSo-PCA) for machinability investigation of graphene nanocomposites, *CIRP Journal of Manufacturing Science and Technology*, 33, 143–157, 2021.
[184] N. Manikandan, R. Raju, D. Palanisamy, J.S. Binoj, Optimisation of spark erosion machining process parameters using hybrid grey relational analysis and artificial neural network model, *International Journal of Machining and Machinability of Materials*, 22(1), 1–23, 2020.
[185] K. Mausam, K. Sharma, P.K. Singh, Aniruddha, Optimization of process productivity for multi phase carbon nanotubes (CNT) reinforced nanocomposites using Taguchi-Fuzzy model, *Advanced Science Letters*, 24(8), 5566–5569, 2018.
[186] P.K. Kharwar, R.K. Verma, Machining performance optimization in drilling of multiwall carbon nano tube/epoxy nanocomposites using GRA-PCA hybrid approach, *Measurement*, 158, 107701, 2020.
[187] R.P. Singh, R. Kataria, J. Kumar, J. Verma, Multi-response optimization of machining characteristics in ultrasonic machining of WC-Co composite through Taguchi method and grey-fuzzy logic, *AIMS Materials Science*, 5(1), 75–92, 2018.
[188] S. Rajesh, D. Devraj, R.S. Pandian, S. Rajakarunakaran, Multi-response optimization of machining parameters on red mud-based aluminum metal matrix composites in turning process, *The International Journal of Advanced Manufacturing Technology*, 67, 811–821, 2013.
[189] N. Neeli, M.P. Jenarthanan, G.D. Kumar, Multi-response optimization for machining GFRP composites using GRA and DFA, *Multidiscipline Modeling in Materials and Structures*, 14(3), 482–496, 2018.
[190] R.K. Verma, K. Abhishek, S. Datta, P.K. Pal, S.S. Mahapatra, Multi-response optimization in machining of GFRP (Epoxy) composites: an integrated approach, *Journal for Manufacturing Science and Production*, 15(3), 267–292, 2015.

[191] B.C. Routar, A.K. Sahoo, A.K. Rout, A.K. Parida, J.R. Behera, Analysis of machining characteristics in drilling of GFRP composite with application of fuzzy logic approach, *International Journal of Industrial Engineering Computations*, 4(4), 447–456, 2013.
[192] P.B. Zaman, S. Saha, N.R. Dhar, Hybrid Taguchi-GRA-PCA approach for multi-response optimisation of turning process parameters under HPC condition, *International Journal of Machining and Machinability of Materials*, 22(3/4), 281–308, 2020.
[193] K. Shunmugesh, K. Panneerselvam, Machinability study of carbon fiber reinforced polymer in the longitudinal and transverse direction and optimization of process parameters using PSO–GSA, *Engineering Science and Technology, an International Journal*, 19(3), 1552–1563, 2016.
[194] G. Talla, D.K. Sahoo, S. Gangopadhyay, C.K. Biswas, Modeling and multi-objective optimization of powder mixed electric discharge machining process of aluminum/alumina metal matrix composite, *Engineering Science and Technology, an International Journal*, 18(3), 369–373, 2015.
[195] B. Das, S. Roy, R.N. Rai, S.C. Saha, Development of an in-situ synthesized multi-component reinforced Al–4.5%Cu–TiC metal matrix composite by FAS technique – Optimization of process parameters, *Engineering Science and Technology, an International Journal*, 19(1), 279–291, 2016.
[196] M.H. Tanvir, A. Hussain, M. M. T. Rahman, S. Ishraq, K. Zishan, SK T.T. Rahul, M.A Habib, Multi-objective optimization of turning operation of stainless steel using a hybrid whale optimization algorithm, *Journal of Manufacturing and Materials Processing*, 4, 64–77, 2020.
[197] A.K. Sahoo, A.N. Baral, A.K. Rout, B.C. Routra, Multi-objective optimization and predictive modeling of surface roughness and material removal rate in turning using grey relational and regression analysis, *Procedia Engineering*, 38, 1606–1627, 2012.
[198] A.K. Sahoo, B. Sahoo, Experimental investigation on flank wear and tool life, cost analysis and mathematical model in turning hardened steel using coated carbide inserts, *International Journal of Industrial Engineering Computations*, 4(4), 571–578, 2013.
[199] R. Kumar, A.K. Sahoo, K. Satyanarayana, G. Rao, Some studies on cutting force and temperature in machining Ti-6Al-4V alloy using regression analysis and ANOVA, *International Journal of Industrial Engineering Computations*, 4(3), 427–436, 2013.
[200] A.K. Sahoo, B. Sahoo, Performance studies of multilayer hard surface coatings (TiN/TiCN/Al2O3/TiN) of indexable carbide inserts in hard machining: part-II (RSM, grey relational and techno economical approach, *Measurement*, 46(8), 2868–2884, 2013.
[201] A.K. Sahoo, B. Sahoo, A comparative study on performance of multilayer coated and uncoated carbide inserts when turning AISI D2 steel under dry environment, *Measurement*, 46(8), 2695–2704, 2013.
[202] D.K. Das, A.K. Sahoo, R. Das, B.C. Routara, Investigations on hard turning using coated carbide insert: grey based Taguchi and regression methodology, *Procedia Materials Science*, 6, 1351–1358, 2014.
[203] D.K. Das, P.C. Mishra, A.K. Sahoo, D. Ghose, Experimental investigation on cutting tool performance during turning AA 6063 using uncoated and multilayer coated carbide inserts, *International Journal of Machining and Machinability of Materials*, 17(3–4), 277–294, 2015.
[204] S.K. Sahu, P.C. Mishra, K. Orra, A.K. Sahoo, Performance assessment in hard turning of AISI 1015 steel under spray impingement cooling and dry environment, *Proceedings of the Institution of Mechanical Engineers, Part B: Journal of Engineering Manufacture*, 229(2), 251–265, 2015.
[205] A. Panda, A.K. Sahoo, R. Rout, Multi-attribute decision making parametric optimization and modeling in hard turning using ceramic insert through grey relational analysis: a case study, *Decision Science Letters*, 5(4), 581–592, 2016.

[206] A. Panda, A.K. Sahoo, A. Rout, Statistical regression modeling and machinability study of hardened AISI 52100 steel using cemented carbide insert, *International Journal of Industrial Engineering Computations*, 8(1), 33–44, 2017.

[207] G.A. Arefi, R. Das. A.K. Sahoo, B.C. Routara, B.K. Nanda, A study on the effect of machining parameters in turning of lead alloy, *Materials Today: Proceedings*, 4(8), 7562–7572, 2017.

[208] A. Panda, A.K. Sahoo, A.K. Rout, Machining performance assessment of hardened AISI 52100 steel using multilayer coated carbide insert, *Journal of Engineering Science and Technology*, 12(6), 1488–1505, 2017.

[209] R. Kumar, A.K. Sahoo, P.C. Mishra, R. Das, M. Ukamanal, Experimental investigation on hard turning using mixed ceramic insert under accelerated cooling environment, *International Journal of Industrial Engineering Computations*, 9(4), 509–522, 2018.

[210] D. Das, S. Mukherjee, S. Dutt, B.B. Nayak, A.K. Sahoo, High speed turning of EN24 steel-a Taguchi based grey relational approach, *Materials Today: Proceedings*, 5(2), 4097–4105, 2018.

[211] R. Tiwari, D. Das, A.K. Sahoo, R. Kumar, R.K. Das, B.C. Routara, Experimental investigation on surface roughness and tool wear in hard turning JIS S45C steel, *Materials Today: Proceedings*, 5(11), 24535–24540, 2018.

[212] A.K. Sahoo, A. Panda, R. Kumar, R.K. Das, D. Das, Investigation on machinability characteristics during turning Al6063 alloy using uncoated carbide insert, *Materials Today: Proceedings*, 5(9), 18120–18128, 2018.

[213] R. Kumar, A.K. Sahoo, P.C. Mishra, R.K. Das, Comparative study on machinability improvement in hard turning using coated and uncoated carbide inserts: part II modeling, multi-response optimization, tool life, and economic aspects, *Advances in Manufacturing*, 6(2), 155–175, 2018.

[214] R. Kumar, A.K. Sahoo, P.C. Mishra, R.K. Das, Performance assessment of air-water and TiO2 nanofluid mist spray cooling during turning hardened AISI D2 steel, *Indian Journal of Engineering and Materials Sciences*, 26(3&4), 235–253, 2019.

[215] R. Kumar, A.K. Sahoo, P.C. Mishra, A. Panda, R.K. Das, S. Roy, Prediction of machining performances in hardened AISI D2 steel, Materials Today: Proceedings, 18, 2486–2495, 2019.

[216] R. Kumar. A.K. Sahoo, P.C. Mishra, R.K. Das, Measurement and machinability study under environmentally conscious spray impingement cooling assisted machining, *Measurement*, 135, 913–927, 2019.

[217] R. Kumar. A.K. Sahoo, P.C. Mishra, R.K. Das, Performance of near dry hard machining through pressurised air water mixture spray impingement cooling environment, *International Journal of Automotive and Mechanical Engineering*, 16(1), 6108–6133, 2019.

[218] R.R. Mishra, A.K. Sahoo, A. Panda, R. Kumar, D. Das, B.C. Routara, MQL machining of high strength steel: a case study on surface quality characteristic, *Materials Today: Proceedings*, 26, 2616–2618, 2020.

[219] R.K. Das, A.K. Sahoo, R. Kumar, S. Roy, P.C. Mishra, T. Mohanty, MQL assisted cleaner machining using PVD TiAlN coated carbide insert: comparative assessment, *Indian Journal of Engineering and Materials Sciences*, 26(5&6), 311–325, 2019.

[220] U. Manoj, P.C. Mishra, A.K. Sahoo, P. Subhashree, Experimental investigation of bio-oil based nanofluid spray cooling during AISI 316 SS turning, in M. Kumar, R. Pandey, V. Kumar (eds) *Advances in Interdisciplinary Engineering*, Lecture Notes in Mechanical Engineering, 277–285. Springer, 2019.

[221] P.H. Nguyen, L.T. Banh, V.D. Bui, D.T. Hoang, Multi-response optimization of process parameters for powder mixed electro-discharge machining according to the surface roughness and surface micro-hardness using Taguchi-TOPSIS, *International Journal of Data and Network Science*, 2, 109–119, 2018.
[222] C. Zhang, Performance of nanocomposite ceramics by wire electrical discharge machining, MATEC Web of Conferences, 34, 01006, 2015.

5 Sustainable Machining of Nanocomposites

CONTENTS

5.1 INTRODUCTION

Sustainability has evolved as an important and strategic priority in manufacturing as its objective is to achieve overall production efficiency considering various dimensions of economic, environmental, and societal aspects. Sustainability assessment plays a very crucial perspective before to implement it in manufacturing industry for cleaner machining. Sustainable manufacturing refers to manufacturing of quality products with minimum cost, i.e., minimal energy/power consumption, etc. A sustainability approach depicts the three important verticals of machining: profitability, ensuring clean/green environment, and enhancing the social relationship between the customer and manufacturer [1]. The economic pillar of sustainable machining denotes maximization of productivity such as material removal rate and tool life and minimization of machining time and tool wear. The environmental pillar implies reduction in power/energy consumption and CO_2 emission during machining because machine tool design and machining processes use 40% and 22% of energy consumption, respectively [2]. The social pillar refers to customer satisfaction by ensuring higher-quality products with minimum dimensional deviation so that the social relationship between customer and manufacturer is enhanced. Thus, surface roughness is considered an important parameter for the societal aspect in sustainable machining. In other words, it is essential to evaluate sustainability assessment for cleaner production in the manufacturing industry not only considering the economical aspect but also society as well as the environment for overall production efficiency. Sustainable manufacturing is referred to as an eco-efficient process, i.e., economically efficient and environmentally safe with social significance that reduces waste and negative impacts and ensures hygienic environment for the worker including worker safety and health, good product quality, waste management, rate of production, training and education, and social relationship with the workers. Thus a sustainable manufacturing process can be summarized as reduction in power/energy consumption; waste reduction; improvements in workers' health; improvement in product quality and

DOI: 10.1201/9781003107743-5

durability; the 3Rs enhancement: recycling, reuse, and re-manufacturing; and the development of renewable energy. The five major aspects should be focused on in the assessment of sustainable machining are energy consumption, machining cost, waste management, health and operator safety, and environmental impact. A case study on the assessment of sustainability during machining is discussed below.

Dash et al. [3] assessed sustainability during machining under various lubrication environments such as dry and nano-fluid minimum quantity lubrication (NFMQL) based on a decision-making approach called Pugh matrix. In the analysis, various sustainability assessment parameters were considered, such as cutting temperature, surface finish, environmental impact, worker safety, coolant cost, coolant recycling and disposal, part cleaning, and noise level for comparison between dry and NFMQL conditions. In the Pugh matrix assessment, some specific weights were allotted to different parameters considering their importance ranging from −2 to +2 for worst-to-best performance and −1 to +1 for worse and better results, respectively. The assignment of scores to various sustainability parameters were as follows:

Environmental impact: As no cutting fluid is used under dry machining, it is environmentally safe and free from any hazardous elements and pollution and thus is allotted the maximum weightage of 2. NFMQL is allotted 1 as a minimum quantity of cutting fluid with nanoparticle is applied with high pressure with little emission of harmful substances. Operator safety: Dry machining produces highly heated chips and machined surface due to the absence of cutting fluids and thus increases worker risk level. Due to this, it is allocated to −2 for dry machining and 2 for NFMQL as it is efficient in cooling/lubrication of chips and their disposal. Coolant cost: As no cutting fluid is used in dry machining, the maximum weightage of 2 is allotted, whereas medium score −1 is allotted to NFMQL due to minimal application of cutting fluids. Surface finish: Improved surface finish is obtained under NFMQL conditions due to good wettability and reduction of friction at the cutting zone compared to dry machining. Thus, the maximum weightage 2 is assigned to NFMQL and −2 for dry condition. Cutting temperature: Due to efficient cooling and lubricating properties of nanofluids under the application of NFMQL, cutting temperature is reduced compared to dry condition. Thus the maximum score of 2 is assigned to NFMQL, whereas lowest score −2 is allotted to dry condition as high cutting temperature is involved because of the absence of cutting fluids. Part cleaning: NFMQL under high pressure easily flushes out all debris from machined part whereas high cost is associated with dry machining for post-part cleaning due to the absence of cutting fluids. As NFMQL does not require any post-part cleaning operation, it is assigned the maximum weightage of 2, whereas the lowest weightage of −2 is assigned to dry machining conditions. Coolant recycling and disposal: As no cutting fluid is applied in dry cutting, the maximum weightage 2 is allotted, whereas 1 is assigned to NFMQL condition due to minimal quantity application of cutting fluids. Noise level: Dry machining induces noise due to friction from the absence of cutting fluid, whereas NFMQL minimizes noise due to efficient lubrication-cooling properties of nanofluids. Thus the maximum score 2 is assigned to NFMQL and the lowest score −2 is allotted to dry condition. After allocation of the scores to various sustainability parameters, the net score of NFMQL, i.e., 9 was much higher than dry machining condition, i.e., −4. It was concluded that machining under NFMQL condition was economical and socio-technologically beneficial.

For sustainable development, minimization of energy consumption followed by lower carbon footprint is of great importance for reduction/savings in manufacturing cost with respect to cost consciousness, as energy crises with environmental problems are a great concern for the manufacturing industry. CO_2 emission also increases with increase in energy consumption. Thus to improve sustainability in machining processes, it is essential to minimize material and energy consumption and pollution from economic and ecological perspectives [3].

5.2 ENVIRONMENTAL ASPECTS

The application of cutting fluids during machining increases cost and environmental impact, not to mention strict protective laws and health regulations putting tremendous pressure on the elimination of cutting fluids in machining. In brief, cutting fluids in machining are associated with economic, environmental, and health concerns. These reasons make dry machining the preferred and best approach to machining. However, the tool wear rate is excessive and surface quality is low under dry machining and thus cannot be applied in many machining applications. In order to overcome this problem and to improve machinability, a new technique has evolved and is gaining popularity nowadays, called environmentally conscious machining, such as minimum quantity lubrication (MQL) or near dry machining (NDM), to replace traditional flood cooling and thus reducing cutting fluid consumption and health hazards substantially. Slowly progress has been made toward the application of environment-friendly nanofluids or nanofluid-assisted MQL machining, hybrid nano green cutting fluids, cryogenic machining, chilled air cooling, and spray impingement cooling to achieve ecological and economic benefits toward sustainable/cleaner/green machining.

The difficult-to-machine materials are broadly classified into three types: hard materials, ductile materials, and nonhomogeneous materials. Because of their three basic properties, i.e., high hardness, high strength, and poor thermal conductivity, these materials are considered difficult-to-cut because machining them leads to less tool life, low productivity, and poor surface quality. Other types of difficult-to-cut materials are polymers and low-carbon steels as they possess high ductility and elongation percentage. Due to poor tool life and surface quality, composites are also called difficult-to-cut materials. The classification and subclassification of difficult-to-cut materials are shown in Figure 5.1 [4]. Various environmentally conscious machining approaches such as dry, MQL, and gas-based coolants like cryogenic and chilled air adopted to eliminate conventional cutting fluids in metal machining are illustrated in Figure 5.2 [4].

Keeping an eye on sustainability, cleaner production, and also strict regulations by The Environmental Protection Agency (EPA), recently new cooling/lubrication strategies have started evolving, such as dry cutting, MQL, cryogenic cooling, and hybrid cooling methodologies, as shown in Figure 5.3 [5].

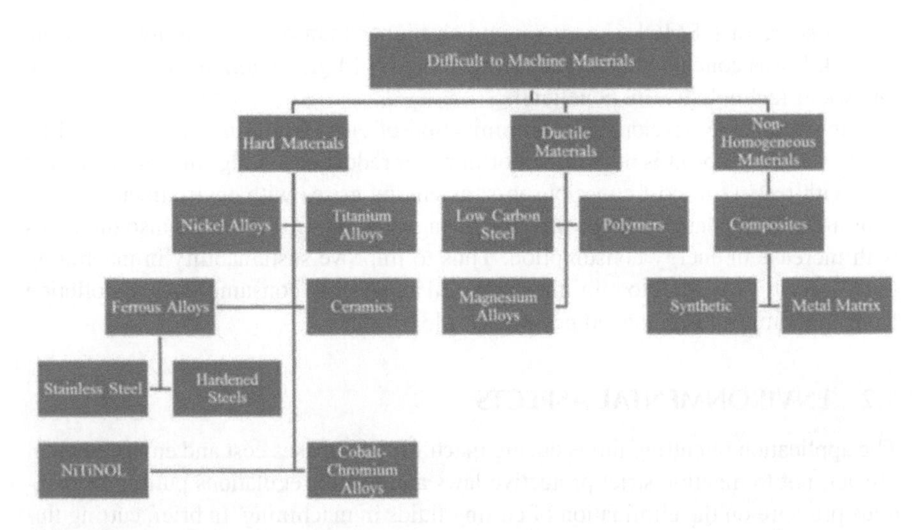

FIGURE 5.1 Classification of difficult-to-machine materials. (From Shokrani, A. et al., *Int. J. Mach. Tools & Manuf.*, 57, 83–101, 2012.)

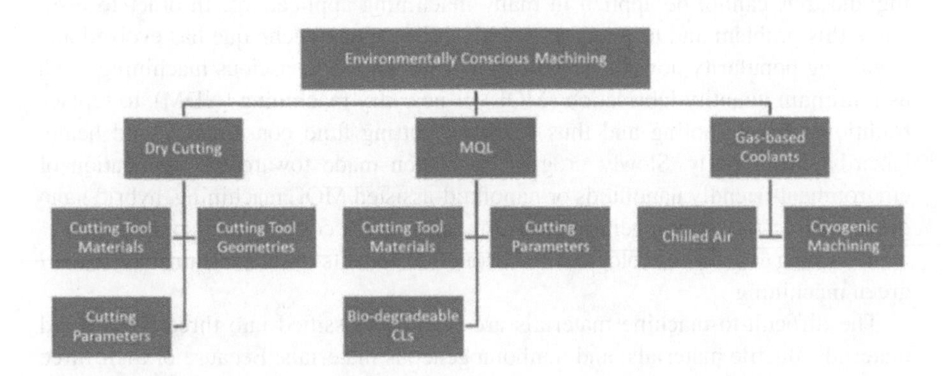

FIGURE 5.2 Classification of different environmentally conscious machining techniques. (From Shokrani, A. et al., *Int. J. Mach. Tools & Manuf.*, 57, 83–101, 2012.)

Nanoparticles mixed with lubricants are also used due to improvements in thermophysical and heat transfer capabilities. This improves machining performance due to reduction in friction and wear. The development of nanofluids for MQL machining applications is also evolving [6].

Many researchers have carried out investigations on sustainable machining considering environmental aspects and have studied machinability using various cooling/lubrication techniques. The outcomes are discussed. Shyam et al. [7] investigated sustainable machining of carbon fiber reinforced silicon carbide ceramic matrix composite under MQL condition considering vegetable-based green cutting fluid and petroleum-based mineral oil. Reduction in surface roughness was noticed to be 17% utilizing green cutting fluid and surface integrity was comparatively better than with

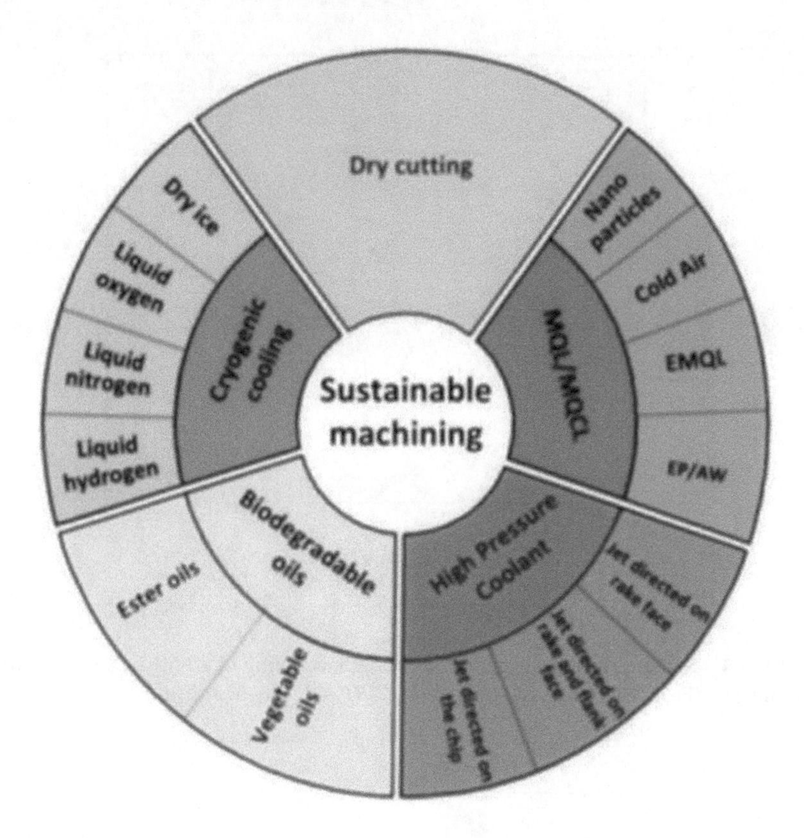

FIGURE 5.3 Ring of sustainable machining. (From Gupta, M.K. et al., *Sustain. Mat. Technol.*, 26, e00218, 2020.)

mineral oil cutting fluid. Alaboodi et al. [8] investigated composite machining with economic, environment-friendly sustainable nanofluids assisted vegetable oil cutting fluids. A comparative machinability assessment under dry machining with coolant oil, sunflower oil + MoS_2, sunflower oil + graphene was carried out. It was observed that the thermal conductivity properties of the renewable cutting fluids increased and thus may replace chemical lubricants during machining composites. Kannan et al. [9] studied the machinability of Al 7075/BN/Al_2O_3 squeeze cast hybrid nanocomposite under different machining environments such as dry and MQL conditions. The effect of feed rate on force, tool wear, and surface roughness was assessed and compared under both environments. Josyula et al. [10] investigated machining of metal matrix composites (MMCs) (Al-5% TiCp) under cryogenic chilled air, liquid nitrogen (LN_2), wet, and dry environments keeping in mind ecological aspects and safety of environment toward sustainable manufacturing. It was observed that LN_2 application performed better in terms of reduction in cutting temperature, surface roughness, and tool wear during machining and improved the machinability of composites, as shown in Figures 5.4–5.6 [10].

Hung et al. [11] studied the effect of cutting fluid on tool wear during machining MMCs. It was revealed that the application of pressurized coolant neither enhances

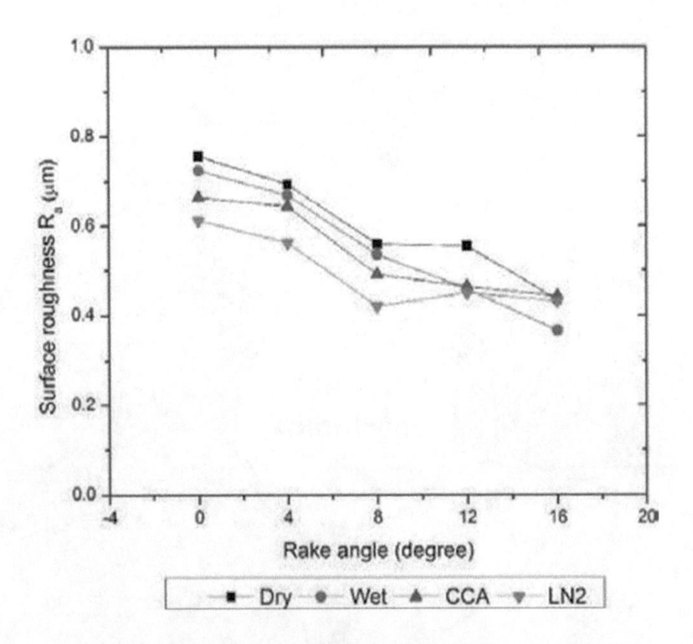

FIGURE 5.4 Surface roughness measured during machining of Al-5%TiC_P under various environments. (From Josyula, S.K. et al., *Proc. CIRP*, 40, 568–573, 2016.)

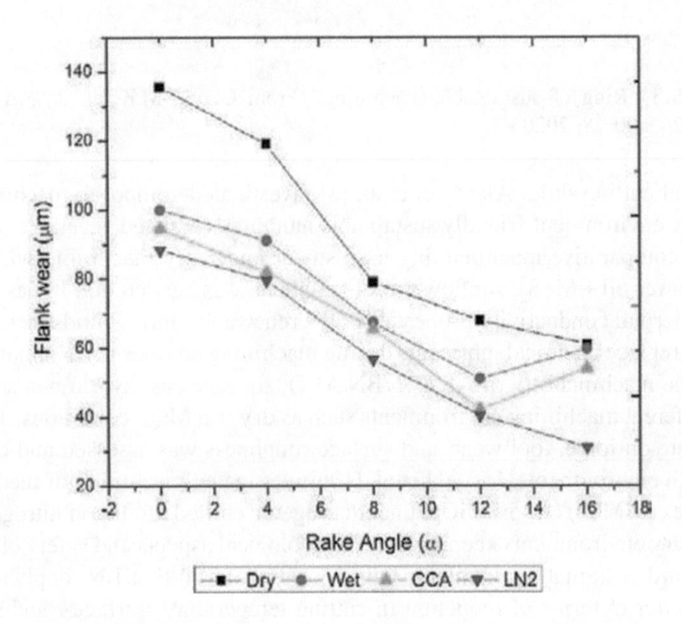

FIGURE 5.5 Flank wear measured under various rake angles while turning under various lubrication strategies. (From Josyula, S.K. et al., *Proc. CIRP*, 40, 568–573, 2016.)

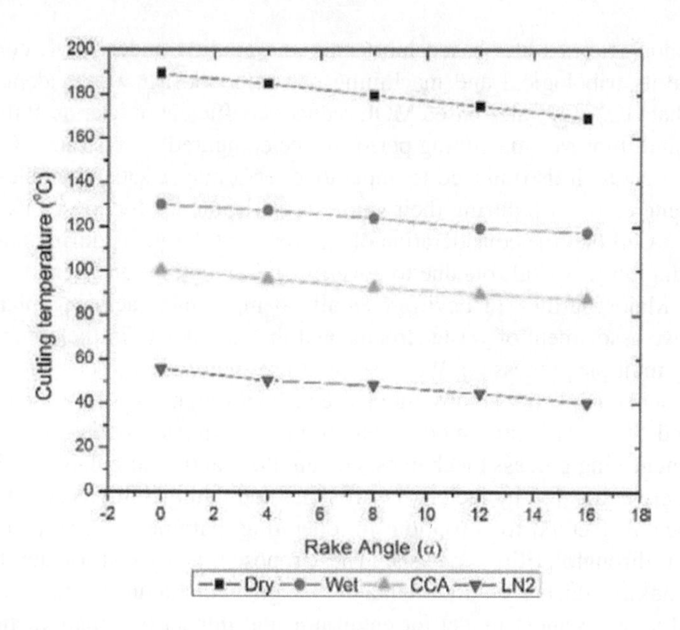

FIGURE 5.6 Cutting temperature with respective rake angle while turning of Al-5% TiC_P under different lubrication strategies. (From Josyula, S.K. et al., *Proc. CIRP*, 40, 568–573, 2016.)

nor decreases the life of the cutting tool. It may be because of inadequate lubrication. Agarwal et al. [12] investigated the influence of machining parameters on surface roughness during drilling of MMCs reinforced with aluminum-based SiC particulate using vegetable oil (Undi oil) applied in an MQL environment. Minimum surface roughness values were obtained for 30% SiC MMCs compared to 10% and 20% for SiCp MMCs. The principal factor influencing surface roughness for MMCs was the percentage of SiCp, and MQL is considered to be an acceptable machining operation due to the use of pollution-free vegetable oil i.e. environmentally friendly. Very minimal amount of oil has been utilized under MQL condition, which is comparatively less than in a flood cooling condition. Sankar et al. [13], on their machining experiment on Al alloy/SiC particulate MMC, concluded that minimum quantity cutting fluid (MQCF) is considered to be eco-friendly as emissions are very less and consumption rate (10 mL/min) is very low compared to flood cooling (400–600 mL/min). MQCF machining outperformed flood cooling as cutting forces, surface roughness, and flank wear were comparatively reduced by 17%, 5%, and 12.5%, respectively. Okokpujje et al. [14] investigated end milling machining of aluminum alloy (AA8112) under different cooling and lubrication environments such as vegetable oil, titanium dioxide (TiO_2), and multi-walled carbon nanotubes (MWCNTs) through quadratic rotatable central composite design of experiment and measured surface roughness. It is observed from the experiment that TiO_2 nano-lubricant under MQL reduces surface roughness of 10% and 17% in comparison to MWCNT nano-lubricant and copra vegetable oil. This shows a cleaner manufacturing of a good-quality product. Lv et al. [15] developed graphene oxide/silicon oxide (GO/SiO_2)

hybrid nanoparticles water-based lubricants and applied under MQL condition to investigate its tribological and machining characteristics. It was evident from the analysis that GO/SiO_2 water-based MQL reduces coefficient of friction at the rubbing interface and improves machining performance compared to separate GO and SiO_2 water-based MQL. It also induced a comparative performance with vegetable-oil-based MQL. Sheng et al. [16], during their study on environmentally conscious manufacturing, assessed that the consideration of environmental aspects during manufacturing planning plays a vital role due to government regulations and demand for green products. Major hurdles in environmentally benign manufacturing planning are comparative assessment of waste streams and the complexity in the computation of evaluating multiple processing. Waste streams are analyzed by a scoring system, and computational complexity is substantially reduced through a feature-based approach. Munoz and Sheng [17] presented an environmental impact model of a machining process integrating process mechanics, coolant flow, and wear behaviors. The waste streams were compared by factors, i.e., toxicity and flammability. The sensitivity of environmental factors to variation in operating parameters are assessed and prioritized through utility analysis. The proposed analytical model helped in decision-making during environmentally benign manufacturing. Choi et al. [18] developed an assessment model for environmental impact in a manufacturing process based on material balance in the process. Solid waste generation, energy consumed, and level of noise were obtained from this analysis to determine if the products were environmentally oriented or not. Jiang et al. [19] developed an environmental evaluation method for a manufacturing process plan where an impact matrix was formed considering different environmental measures for various operations. Then score for each measure was obtained by assigning weights from the analytical hierarchy process (AHP) and provides a decision on environmental impact of manufacturing. Avram et al. [20] developed a sustainability score utilizing the weighted summation method to obtain the weights of multi-criteria for machining tool systems through AHP. Tan et al. [21] established a model for decision-making on the selection of cutting fluids considering quality, cost, and environmental impact taken together to achieve green manufacturing. The optimal use of cutting fluid minimizes environmental pollution during machining, minimizes cost, and maximizes product quality. Yan et al. [22] assessed the sustainability of machining processes considering three important criteria: environmental, economic, and societal. The weights of the different sustainability criteria were evaluated using entropy weight approach, and sustainability performance was then obtained using extension theory. In order to define the membership degree of each machining parameter, a matter element model and a correlation function are established for sustainability assessment. Mia et al. [23] investigated hard machining under different cooling-lubrication sustainable environments such as dry, MQL, and solid lubricants with compressed air (SL+ CA) and measured surface roughness, tool wear, cutting temperature, and chip characteristics. Furthermore, sustainability assessment has been established through the Pugh matrix environmental approach considering parameters such as environmental impact, operator health, cost of coolant, coolant recycling and disposal cost, part cleaning, and machining outputs. It is evident that the MQL system improves machinability during machining and enables environmentally friendly cleaner machining. Dash et al. [24] investigated surface integrity and chip morphology in machining hardened steel under nanofluid MQL cooling environment and also assessed economic and sustainability aspects through the decision-making tool Pugh matrix. Various sustainability assessment parameters were considered in the study such as worker safety, cutting temperature, surface roughness, environmental impact, coolant cost, recycling and disposal cost of coolant, part cleaning, and noise level, and scores were assigned to parameters depending upon their importance, ranging from −2 to +2 (inferior to superior results). It is evident from the sustainability analysis and also the from Kiviat radar diagram that machining with NFMQL is socio-technologically effective as well as economically viable. Padhan et al. [25] investigated machinability and sustainability assessment during machining of austenitic stainless steel under various cooling and lubrication environments such as dry, compressed air (CA), flooded, and MQL. The Pugh environmental sustainability approach was utilized to assess sustainability. The MQL environment was considered as an environmentally friendly, cleaner machining and improved the sustainability during finish machining. Panda et al. [26] studied the machinability of AISI D3 steel using ceramic insert in terms of cutting force, tool wear, and surface roughness by varying process parameters such as approach angle, nose radius, cutting speed, feed rate, and depth of cut under a dry environment. Prediction models were developed by multiple regression analysis and their adequacy was checked. Multi-response parametric optimization was been carried out using hybrid techniques such as RSM-GA-PSO (response surface methodology, genetic algorithm, particle swarm optimization) and validated by confirmation runs. Optimal parameters have been used to estimate energy consumption and savings in carbon footprint. Finally, sustainability assessment, economic analysis, and energy savings were analyzed by Pugh matrix, Gilbert's approach, and carbon footprint analysis for cleaner/green machining. Abbas [27] investigated machining performance of AISI 1040 steel under MQL nanofluid (vegetable oil mixed with Al_2O_3 nanoparticles), dry, and flood environments and obtained optimal parameters, which cover both machining outputs and sustainability (carbon dioxide emission and machining cost) aspects. MQL nanofluid outperformed dry and flood cooling conditions as the best surface quality is obtained due to improved frictional behavior of the MQL nanofluid mist at the interface. Use of MQL nanofluid improved surface quality by 34.5% and 85.5% compared to flood cooling and dry cutting. MQL nanofluid outperformed dry and flood cooling conditions as lowest power consumption was observed. MQL may be considered as an effective sustainable method as power consumption and CO_2 emission are substantially reduced. Dry cutting offers less total machining cost compared to the other two cooling techniques due to elimination of cost related to cutting fluids and nanofluids and can be considered an effective, sustainable approach. The lowest machining cost has been obtained in test no. 7 (cutting speed of 150 m/min, depth of cut of 0.25 mm, and feed rate of 0.18 mm/rev) for all cutting environments. Dry cutting reduces machining cost by 7.9% compared to MQL nanofluid application. Abbas et al. [28] investigated sustainability assessment during machining under nanofluid MQL, dry, and flood cooling with measured responses such as surface roughness and power consumption along with impact on environment, cost of machining, waste management, and safety and health of operators. Sustainability assessment results for dry, flood, MQL nanofluid tests were analyzed along with a comparison between all cooling-lubrication based on weighted sustainability index (TWSI) results. For dry and flood cutting test, test no. 9 (cutting speed of 150 m/min, feed of 0.06 mm/rev, and depth of cut of 0.25 mm) was found to be an optimal sustainable run based on highest TWSI. However, dry cutting offers higher TWSI compared to flood cooling at the same cutting conditions in test no. 9. Test no. 21 (cutting speed of 100 m/min, feed rate of 0.06 mm/rev, and depth of cut of 0.75 mm) was found to be an optimal sustainable run based on highest TWSI for the MQL nanofluid cutting test [28]. Comparisons between most sustainable run for dry, flood, and MQL nanofluid have been performed. The performance of nanofluid MQL (Al_2O_3 particle suspended at oil–water mixture spray) was observed to be more sustainable with a total TWSI of 0.7 during machining followed by dry machining of 0.52 and flood coolant of 0.4 [28]. Machinability of MQL nanofluid has been improved due to effective heat transfer and tribological action by the nanoparticle mist that reduces surface roughness and power consumption for sustainability. The optimal parametric settings are found to be cutting speed of 116 m/min, feed rate of 0.06 mm/rev, and depth of cut of 0.25 mm with the highest desirability at 0.9050. Minimum surface roughness and power consumption at optimal settings were 0.354 μm and 0.528 kW, respectively. Gajrani et al. [29] investigated machining performance of hard materials through minimum quantity cutting fluids using vegetable-based green cutting fluids with additives such as solid lubricant nanoparticles. It was observed that a 0.3% concentration of solid lubricant MoS_2-based hybrid nano green cutting fluid outperformed other cutting fluids such as mineral oil, green cutting fluids, 0.3% concentration of CaF_2-based hybrid nano green cutting fluid with respect to surface roughness (37% better surface finish), cutting force (17% reduction), feed force (28% reduction), and coefficient of friction at chip-tool interface (11% reduction). Sreejith [30] conducted machining of 6061 aluminum alloy and studied the influence of different cutting environments such as dry, MQL, and flooded coolant on cutting force, surface roughness, and tool wear. It is evident that machinability under MQL performed better compared to dry and flood applications and can be considered as an environmentally friendly, cleaner machining. Masoudi et al. [31], on their machining operation of AISI 1045 steel, concluded that MQL cutting condition improved the performance compared to dry and wet cooling with respect to surface topography, cylindricity, and cutting force. Sustainability assessment using the Pugh matrix approach considering environmental impact, operator health, economy, and production efficiency revealed superior performance of MQL machining over dry and wet machining. Goldberg [32] recommended helical cutting edge geometry and variable pitch flute pattern to reduce vibration during machining. Productivity improved a lot utilizing the proposed tooling system along with dry or MQL conditions with ecological benefits and minimization of energy consumption. Upadhyay et al. [33] identified the problems associated with flood cooling application in machining and the need for MQL followed by its working principle. Based on the review on turning and milling, the influence of operating parameters on MQL performance and the effect of MQL on machinability aspects have been presented. Faga et al. [34] investigated the influence of dry, near dry such as MQL environment during machining of titanium alloy for sustainability implication considering tool wear, cutting forces, surface quality, lubricant consumption, and health hazard. The study also proposes environmental impact minimization along with productivity enhancement. Shokoohi et al. [35] developed an eco-friendly water-mixed vegetable oil cutting fluid integrated with an antibacterial agent and applied it through MQL with new cooling approach, i.e., pre-cooling the workpiece, during machining of hardened AISI 1045 steel. Machining performance was assessed through responses such as surface roughness, power consumption, and chip formation, including machining hazards. Significant improvement was observed in health and ecological aspects as well as performance in machining along with control of bacterial growth with application of vegetable oil compared to straight oil. Debnath et al. [36] reviewed the development of vegetable oil bio-based cutting fluids and their application in machining and observed the minimization of ecological issues. A brief technique on dry cutting, MQL and cryogenic cooling cleaner approach in machining has also been presented. A significant reduction in cutting fluids with better machining performance was also noticed compared to the conventional wet cooling approach. Deiab et al. [37] recommended a better machining performance of MQL and minimum quantity cooled lubrication (MQCL) environment with rapeseed vegetable oil cutting fluid for sustainability, compared to dry and cryogenic cooling, and provided better results for tool wear, surface roughness, and energy consumption, as shown in Figures 5.7–5.9 [37]. Vegetable cutting fluid along with environmentally benign cooling techniques such as MQL and MQCL seems to be a sustainable alternative over conventional coolants for machining.

Priarone et al. [38] studied the application of emulsion mist cutting fluids during machining of Ti-48Al-2Cr-2Nb alloy, and noticed improved tool life compared to dry and MQL cutting. Jamil et al. [39] studied the sustainability measures (specific cutting energy, carbon emissions, energy efficiency, and process time) and machinability

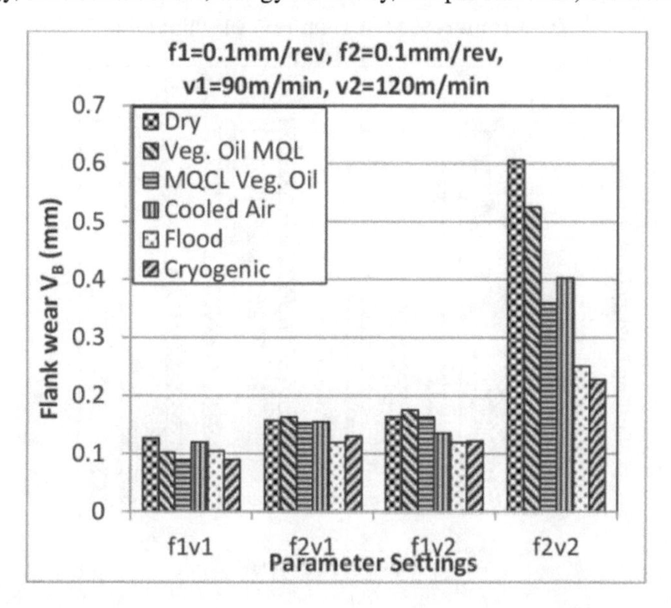

FIGURE 5.7 Tool wear with lubrication techniques. (From Deiab, I. et al., *Proc. CIRP*, 17, 766–771, 2014.)

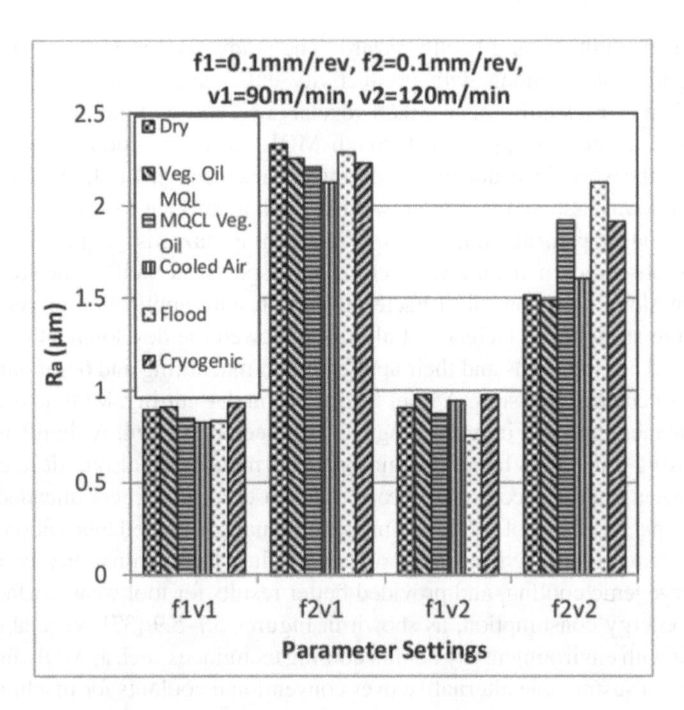

FIGURE 5.8 Surface roughness with lubrication techniques. (From Deiab, I. et al., *Proc. CIRP*, 17, 766–771, 2014.)

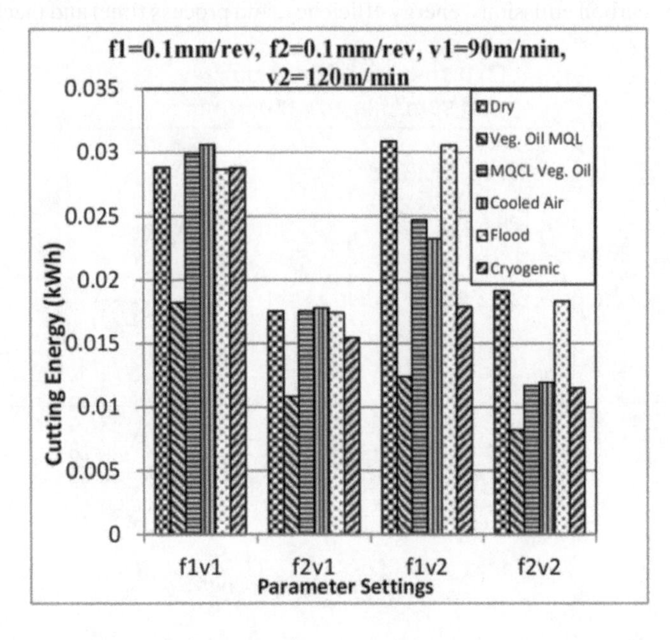

FIGURE 5.9 Cutting energy consumption with lubrication techniques. (From Deiab, I. et al., *Proc. CIRP*, 17, 766–771, 2014.)

during milling of titanium alloy through different cooling and lubrication techniques, such as MQL, CO_2-snow, cryogenic LN_2, and dry cutting and measured responses such as surface roughness, tool wear, and cutting temperature. Based on the investigation, CO_2-snow outperformed in terms of sustainability and machinability compared to cryogenic-LN_2, MQL, and dry cutting, as shown in Figures 5.10–5.12 [39].

Gupta et al. [40] investigated sustainable machining of Al 7075-T6 alloy under different cooling-lubrication techniques such as dry, nitrogen cooling, nitrogen MQL, and Ranque–Hilsch vertex tube nitrogen MQL (R-N_2 MQL). It is evident that R-N_2 MQL performed better compared to other cooling methods as reduction in surface roughness and tool wear were found to be 77% and 118%, respectively, thus enhancing sustainability by saving resources. Pereira et al. [41] investigated machining of Inconel 718 using natural, biodegradable-oil-assisted MQL and compared with other oils such as sunflower oil, oleic sunflower oil, castor oil, and ECO-350 recycled oil. ECO-350 recycled oil was observed to be feasible and increased cutting tool life by 30% compared to canola oil, but the recycling process needed improvement. Considering all issues, oleic sunflower oil improves tool life by 15% with environmental impacts similar to canola oil and thus may be considered an efficient, environmentally friendly, and technically viable. Kumar et al. [42] conducted machining of hardened AISI D2 steel under environmentally conscious conditions such as spray impingement cooling to investigate machinability characteristics by measuring tool wear, cutting temperature, and surface roughness. Flank wear and surface roughness were well within the recommended limit of 0.2 mm and 1.6 μm, respectively, in all the runs, which shows the benefit of the application of air-water spray cooling in machining. The images of flank wear at respective experimental runs as per the Taguchi L16 orthogonal array design of experiment are shown in Figure 5.13 [42].

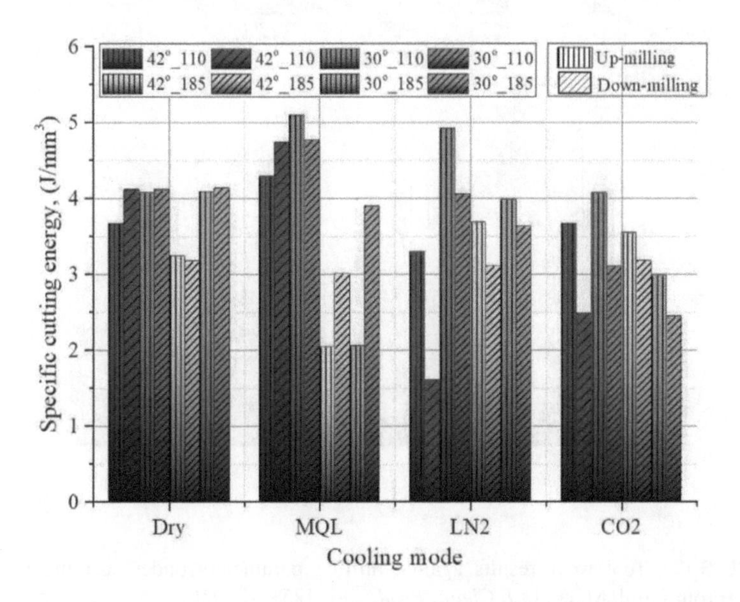

FIGURE 5.10 Specific cutting energy at varying cutting speeds, helix angles, and cutting environments. (From Jamil, M. et al., *J. Clean. Prod.*, 281, 125374, 2021.)

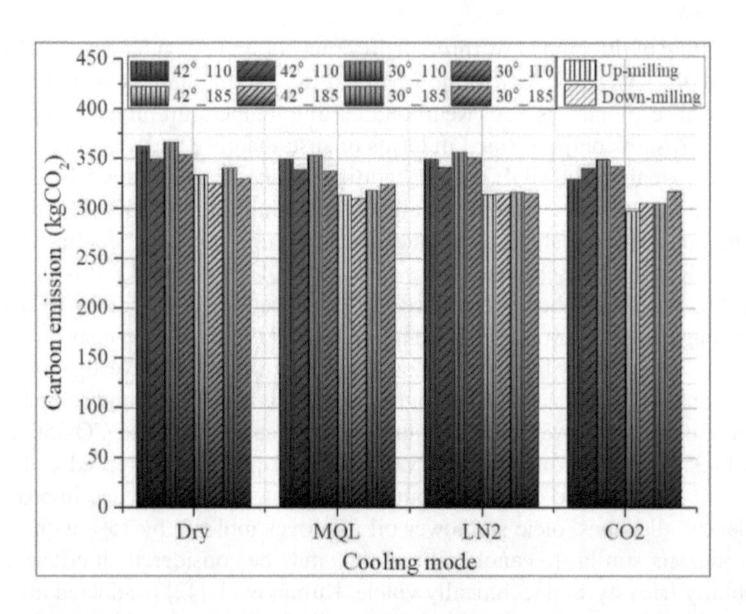

FIGURE 5.11 Carbon emission (kg-CO_2) against milling parameters and coolant modes. (From Jamil, M. et al., *J. Clean. Prod.*, 281, 125374, 2021.)

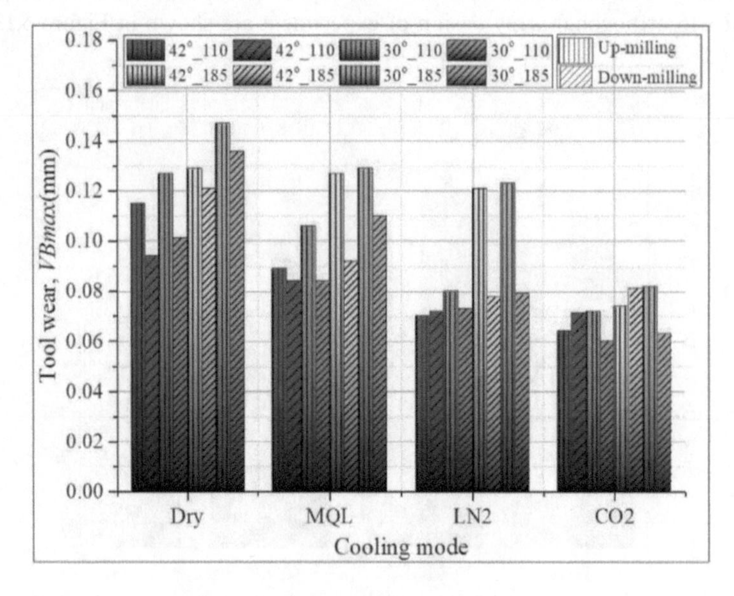

FIGURE 5.12 Tool wear results against milling parameters under sustainable cooling modes. (From Jamil, M. et al., *J. Clean. Prod.*, 281, 125374, 2021.)

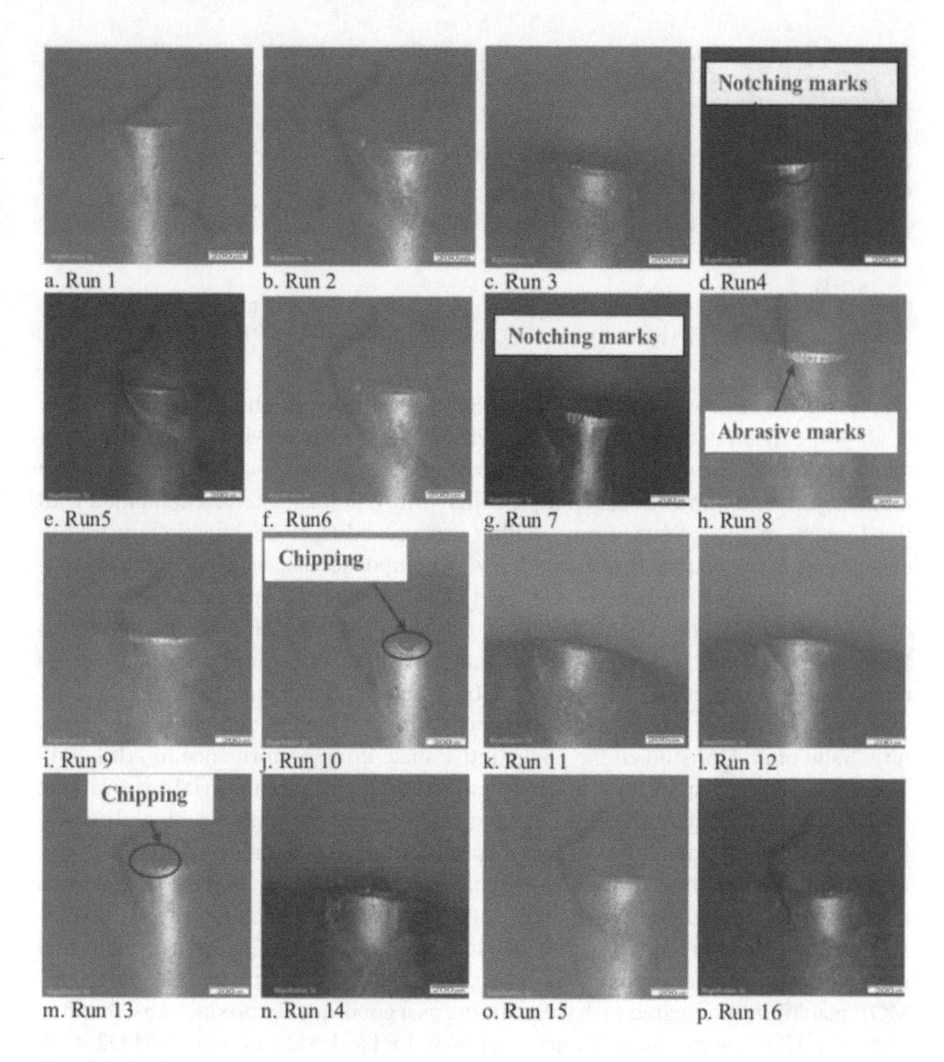

FIGURE 5.13 (a–p) Flank wear images of respective experimental runs. (From Kumar, R. et al., *Measurement*, 135, 913–927, 2019.)

The proposed spray cooling technique reduces the cutting temperature because the latent heat is being absorbed by the water droplets through evaporation, thus slowing down tool wear and improving tool life and surface quality. Cutting temperature was between 120.4°C and 217°C. The images of cutting temperatures for Runs 1–4 are shown in Figure 5.14 [42]. From an industrial application point of view, it may be considered as a stable, environmentally and ecologically cleaner machining of hardened AISI D2 steel using multilayer coated carbide inserts.

Kumar et al. [43], in another investigation on machinability of AISI D2 steel, applied spray cooling to study chip reduction coefficient and flank wear and developed artificial neural network (ANN) prediction models using the feed-forward back propagation algorithm. The model was found to be accurate and effective in

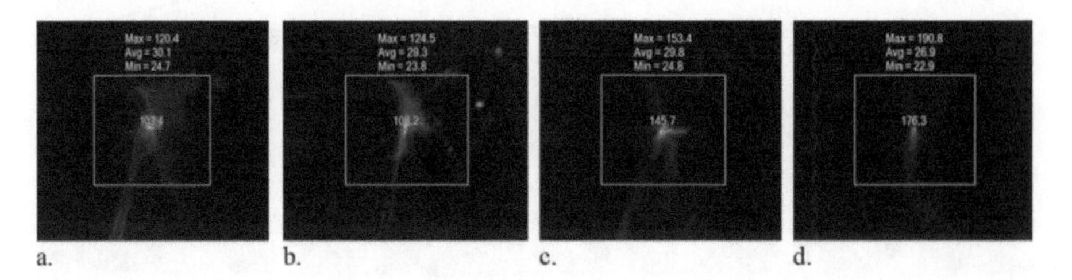

FIGURE 5.14 Image of cutting temperature using infrared camera at (a) Run 1, (b) Run 2, (c) Run 3, and (d) Run 4. (From Kumar, R. et al., *Measurement*, 135, 913–927, 2019.)

predicting responses as its R^2 value is high with minimum absolute error. Mishra et al. [44] carried out machining of AA 7075/SiC particulate MMCs under dry and spray cooling environments and optimized the machining parameters through a multi-objective optimization technique called gray relational analysis combined with the Taguchi method. Water spray cooling was proved to be beneficial compared to dry cutting based on machinability studies of composite materials and was also considered environmentally benign. Kumar et al. [45] performed sustainable machining of AISI D2 steel using coated carbide insert under dry environment and developed RSM and ANN prediction models for flank wear, surface roughness, and cutting temperature. The accuracy of the ANN model was better compared to RSM for flank wear, whereas RSM model was better for surface roughness and cutting temperature. Sahu et al. [46] studied the comparative machining performance of AISI 1045 steel under dry and spray cooling environments by measuring material removal rate (productivity), cutting temperature, and surface roughness. The machinability performance of environmentally conscious spray cooling was improved than that of dry machining. The optimized results yield maximization of material removal rate with minimum surface roughness and temperature. Das et al. [47] conducted machinability investigation of hardened AISI 4340 steel under dry and MQL environments and measured surface roughness and tool wear. Satisfactory performance was noticed for MQL machining compared to dry machining as a good-quality product was obtained under the MQL environment. Kumar et al. [48], for hard machining of AISI D2 steel, considered nanofluids (Al_2O_3 and TiO_2 water based) at different concentrations of weight and applied them to the cutting zone through air-assisted spray impingement cooling using an inexpensive coated carbide insert. Very limited investigation has been noticed during hard machining under nanofluids-assisted air-water spray cooling. Machinability performance was observed to be better using TiO_2 nanofluid compared to Al_2O_3 nanofluid due to better lubrication, wettability, and heat dissipation characteristics, particularly at a 0.01% weight concentration of TiO_2. At this concentration, there was 29%, 9.7%, and 14.3% reduction in tool wear, cutting temperature, and surface roughness compared to Al_2O_3 nanofluid. Also there is reduction of 29% and 27.7% in tool wear using 0.01% weight concentration of TiO_2 compared to dry and air-water spray cooling. Tool life at 0.01% concentration of TiO_2 nanofluid is 2.52 and 1.47 times higher than dry cutting and air-water spray cooling, respectively. Roy et al. [49] reviewed different cooling/lubrication strategies for MQL machining as well as various cutting fluids such as vegetable oil, mineral oil, synthetic oil, and a nanofluid-assisted environment, particularly in sustainable machining applications aiming at environmental, economic, and technological benefits. Das et al. [50] investigated comparative machinability performance (tool wear and surface roughness) of hardened AISI 4340 steel under dry and MQL environments. MQL machining was observed to be better compared even at higher cutting speeds. Ukamanal et al. [51] developed TiO_2 nanoparticle from a high-energy ball milling process and applied it in the machining of AISI 316 stainless steel as a water-based spray cooling. From this experiment, it was observed that at 0.03% weight concentration of nanofluid, tool and chip temperature were found was low. Ukamanal et al. [52] optimized machining and spray cooling parameters for multi-responses (tool temperature, chip temperature, surface roughness, and flank wear) during machining of AISI 316 stainless steel through weighted principal component analysis coupled with the Taguchi method. Machining under spray cooling was found to be effective compared to dry machining as a substantial reduction in cutting temperature was noticed. Dambhare et al. [53] investigated the sustainability aspects of turning with a case study in the Indian machining industry. The effect of process parameters, cutting environments, and type of cutting tools on sustainability factors such as surface roughness, material removal rate, and energy consumption was studied. From an ANOVA, it was revealed that cutting environment and type of tool affected surface roughness. Material removal rate was influenced majorly by cutting parameters and type of tool, whereas energy consumption was influenced by cutting environment, type of tool, cutting velocity, and depth of cut. Prediction models through RSM and experimental results are very close to each other. The parameters are optimized for sustainability with due weightage/importance to minimum power consumption through the desirability approach. Das et al. [54] investigated machining of hardened AISI 4340 steel under various cooling applications such as CA, water-soluble-coolant-assisted MQL, and a nanofluid (Al_2O_3 nanoparticle with eco-friendly radiator coolant)-assisted MQL. The nanofluid-assisted MQL application showed better machining performance (reduction in flank wear and cutting force, and improved surface finish) compared to the other two. The MQL technique was observed to be effective in minimizing health risk and machining cost. Iskandar et al. [55] studied MQL machining, i.e., a slotting test of carbon-fiber-reinforced polymer laminate composites to assess machinability, compared to dry and flood cooling. High air flow rate and low oil flow rate induced higher tool life with minimum machining error, which was in agreement with flow visualization analysis. Under MQL conditions, reduction in flank wear was noticed at 30% and 22% compared to pressurized air, dry, and flood coolant. Haq et al. [56] investigated face milling operation of IN718 (Inconel 718) under two environmentally conscious water-based lubricating conditions such as MQL and NFMQL and measured surface roughness, temperature, power consumption, and material removal rate. Superior sustainability rating has been observed in NFMQL cutting condition with an achievable surface roughness of 0.1 μm or even lower. Qu et al. [57] studied the grinding of carbon-fiber-reinforced ceramic matrix composites (C_f/SiC) using different environmental conditions such as dry, flood, MQL, and carbon nanofluid MQL. The influence of lubrication conditions and lubrication parameters on grinding characteristics was investigated. The study clearly showed the application potential of carbon nanofluid MQL for efficient lubrication and proposed the greener/cleaner manufacturing method. Giasin et al. [58] investigated drilling performance of glass-aluminum-reinforced epoxy (GLARE) fiber-metal laminates under MQL and cryogenic liquid nitrogen environments and studied surface roughness, cutting forces, condition of cutting tool, and micro-hardness after machining. Both cooling conditions were found to be effective in drilling of laminates by providing improved surface finish of workpiece and yielding environment-friendly machining. Giasin et al. [59] studied the drilling of GLARE 2B fiber-metal laminates under MQL and cryogenic liquid nitrogen environments to study burr formation, deviation in hole size, and circularity. A 70% reduction in hole circularity was observed under liquid nitrogen environment compared to MQL and dry conditions. Also, exit burr formation was reduced significantly under cryogenic liquid nitrogen and MQL environments.

5.3 SOCIAL ASPECTS

The basic focus of social sustainability is on health improvement, safety of operator, enhancement of quality of life of people, ethics, training and education, and social relationship with workers. It broadly covers the following: (a) Worker health particularly at the shop floor and supervisory personnel and its direct impact on manufacturing processes. The emissions and wastes coming out from the machining operation directly affects labor. (b) Worker safety, i.e., following safety precautions during machining operations to avoid any unforeseen accidents and ensuring personnel safety by implementing regulatory safety requirements. (c) Labor relations mainly wages, working hours, work load, etc. play an important role in making the organization socially sustainable. (d) Training and education on a regular basis to upgrade the skills of workers and managers that help in improving the quality of work life and thus enhance organizational sustainability [60]. Some research findings related to the social aspects are described next.

Bhanot et al. [60] studied the optimization of sustainable parameters (production cost, part quality, environmental impact, worker health and safety, etc.) for competitive manufacturing as there should be a balance between the three pillars of economic, environmental, and societal verticals. Social dimensions considered for the study are worker health, worker safety, labor relations, and training and education. Sustainable manufacturing parameters are divided into two groups based on modularity analysis. Group I consists of PM, PC, PR, WS, LR, and TR whereas Group II consists of CQ, ML, EI, WH, ER, WI, and WM. It can be inferred that Group I concerns with the interdependency between social and economic parameters and Group II concerns with the impact of environmental parameters on workers and machining quality, as shown Figure 5.15 [60].

The study proposes a graph-assisted expert system for the analysis of the outcome of various factor changes, which has been obtained by a model based on interrelationships among the different parameters. Considering two parameter groups, it helps in designing the manufacturing processes. Tamang et al. [1] carried out sustainable machining of Inconel 825 under dry and MQL environments. A significant improvement in tool wear, power consumption, and surface finish Was obtained under the MQL study. MQL shows a significant reduction in tool wear and power

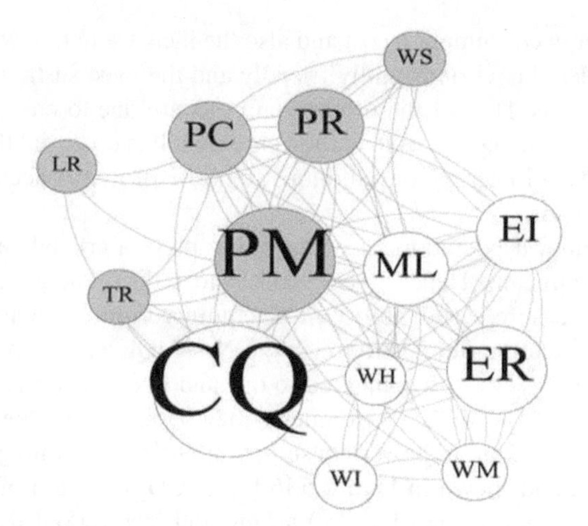

FIGURE 5.15 Relationship between sustainable manufacturing parameters. Group I consists of PM (Process management), PC (Production cost), PR (Production rate), WS (Worker safety), LR (Labor relations), and TR (Training and education), whereas Group II consists of CQ (Cutting quality), ML (Materials), EI (Energy intensity), WH (Worker health), ER (Environmental regulations), WI (Water intensity), and WM (Waste management) (From Bhanot, N. et al., *Proced.- Soci. Behav. Sci.*, 189, 57–63, 2015.)

consumption. Surface roughness values obtained under MQL condition were found to be low compared to dry machining environments. For sustainability of economic, social, and environmental aspects, machining parameters are optimized, i.e., minimization of power, surface roughness, and flank wear simultaneously through genetic algorithm (GA) and a 16 Pareto-optimal fronts solution. The optimal solution showed better convergent capability for wide application in machining industries. This solution also provides the optimal combination of machining parameters to yield minimal responses. From investigation, MQL was found to be effective with respect to operator health and environmental consciousness for cleaner machining. Gupta et al. [5] investigated sustainable machining of titanium alloy under dry, liquid nitrogen, and hybrid liquid nitrogen with MQL (LN_2+MQL) environments to study machinability characteristics (surface roughness, cutting force, and temperature) and environmental parameter impact (cycle time, productivity, energy consumption, carbon emission, and economic analysis). Sustainability assessment has also been performed through AHP coupled with TOPSIS. The hybrid liquid nitrogen with MQL condition outperformed and minimized machinability responses and environmental indices and improved productivity by 34.21% compared to dry cutting. A noticeable difference in total machining time (productivity) was observed under various cooling/lubrication techniques. Increase in total cycle time reduces process productivity. Due to the reduction in time required to replace the tool and overall machining time, hybrid liquid nitrogen with MQL condition improved tool life by delaying tool wear. The hybrid liquid nitrogen with MQL (LN_2+MQL) environment reduced total process cost by up to 65.84% compared to dry and LN_2 cooling. The best result occurred for LN_2 and LN_2+MQL cooling due to the reduction of major costs such as tool cost and energy consumption cost and also the increase in tool wear. LN_2+MQL cooling is considered environmentally friendly and the most sustainable alternative over other conditions. Half of the production costs are due to energy consumption and environmental damages dues to carbon emission. It is observed that [5], industry shares around 31% of energy consumption, and 60% of that is accountable for the manufacturing sector.

Thus, chip removal operation, i.e., machining plays a crucial role in manufacturing and is significant. Hence it is essential to reduce energy consumption in machine tools to ensure sustainability in machining. From comparative investigations at different cooling/lubrication cases, LN_2+MQL environments consumed 15.89% and 3.07% less energy compared to dry and LN_2 machining. The favorable result from hybrid LN_2+MQL environments may be attributed due to reduction in friction and thus reduction in power consumption during machining. CO_2 emission result is observed and shown in Figure 5.16 [5]. A CO_2 emission of 10.71 kg under dry conditions at a cutting speed of 150 m/min and feed rate of 0.15 mm/rev was reduced to 6.58 kg using a hybrid LN_2+MQL environment at a cutting speed of 100 m/min and feed rate of 0.1 mm/rev. Thus, a 38.56% reduction in CO_2 emission was achieved at optimum processing conditions [5]. The surface roughness obtained for the LN_2+MQL condition is comparatively less than in dry and LN_2 machining. It was observed that the LN_2+MQL condition produces 32.84% and 20.34% lower surface roughness compared to dry and LN_2 machining environments. Surface quality was improved with increase in cutting speed and reduced with rise in feed rate. The cutting temperature obtained for the LN_2+MQL condition was comparatively less than dry and LN_2 machining. The produced cutting temperature for the LN_2+MQL

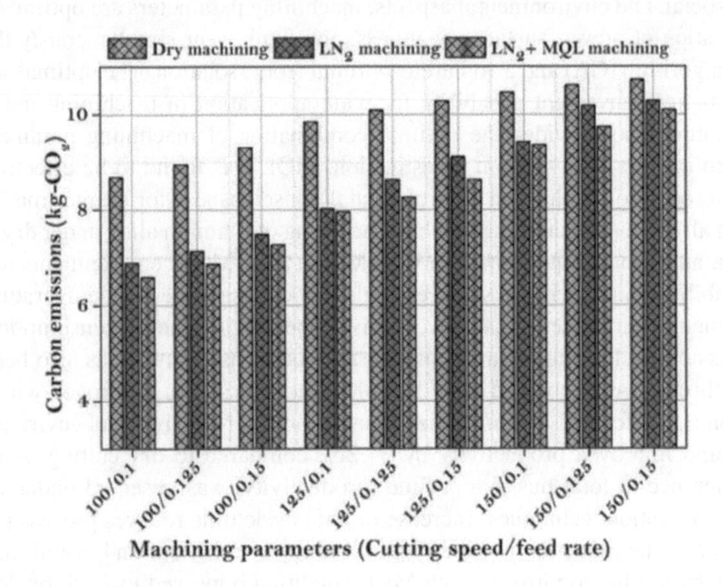

FIGURE 5.16 Carbon emission values under dry, LN_2, and LN_2+MQL environments at different experimental runs. (From Gupta, M.K. et al., *Sustain. Mat. Technol.*, 26, e00218, 2020.)

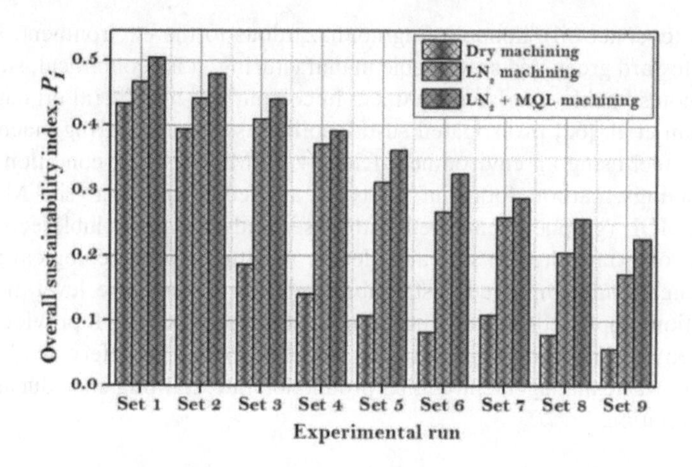

FIGURE 5.17 Overall sustainability index and sustainability assessment results under dry, LN_2, and LN_2+MQL machining. (From Gupta, M.K. et al., *Sustain. Mat. Technol.*, 26, e00218, 2020.)

condition was 43.77% and 40.16% lower than dry machining. It may be attributed due to the heat absorption of cryogenic LN_2+MQL cooling, which cools the workpiece and tool surface. In addition to that, MQL reduces friction by lubricating the contact surface. Cutting temperature increases with increase in cutting speed and decreases with rise in feed rate. The cutting force obtained for the LN_2+MQL condition is comparatively less than for dry and LN_2 machining. In general, dry machining yields 27.97% and 36.67% higher cutting force than LN_2 and LN_2+MQL environments, respectively. The reason for a less cutting force under cryogenic environments may be attributed to the reduction in cutting temperature, tool wear, built-up-edge (BUE) formation, and reduction in friction by MQL lubrication. Cutting force reduces with increase in cutting speed and increases with increase in feed rate. Machining under LN_2+MQL environments has been improved compared to other cooling conditions as per sustainability assessment and higher overall sustainability index, as shown in Figure 5.17 [5].

Also, a higher sustainability index has been observed for experiment 1 (cutting speed of 100 m/min, feed rate of 0.1 mm/rev, and depth of cut of 0.75 mm) compared to other conditions and a lower index is seen in experiment 9 (cutting speed of 150 m/min, feed rate of 0.15 mm/rev, and depth of cut of 0.75 mm). From the analysis, it is evident that machinability performance is improved under LN_2+MQL environments and may be considered as an environmentally friendly cooling condition from a sustainability perspective compared to other cooling conditions [5]. Somashekaraiah et al. [61] compared the performance of a developed green cutting fluid (GCF) over commercial metalworking fluid (COM) during sustainable machining. Machinability performance of mild steel was found to be better with GCF compared to mineral oil as reduction in cutting force was observed. GCF was found to prevent corrosion and Gram-negative bacteria growth compared to mineral oil that prevents environmental hazards and has better antimicrobial properties. Renewable GCF is nontoxic, biodegradable, contains eco-friendly base materials (coconut oil), nonpolluting/nonhazardous to the environment, and thus effective toward green and sustainable manufacturing. It is economical, stable, and also promotes healthy work environment in comparison to mineral oil cutting fluids. Padhan et al. [62] investigated sustainability assessment during machining of hardened steel using an environment-friendly NFMQL cutting condition through energy savings, carbon footprint analysis, and economic analysis. Machining under NFMQL (graphene enriched with base fluid as water-soluble eco-friendly radiator coolant) resulted in safe and cleaner machining with ecological and economic benefits and improved sustainability due to lower noise level and power consumption, improved surface finish, and better worker safety. It provides a more hygienic environment for workers along with environmental safety, worker safety and health, waste management, higher production rate, training and education, and workers' relation.

5.4 ECONOMICAL ASPECTS

In manufacturing processes, machining cost plays a vital role for the economy and focuses on direct and indirect costs for capital, environmental, and societal factors. It has the greatest impact on economy and technological viability. The various economic parametric requirements for manufacturing are as follows: (a) production cost, which is to be kept as minimum as possible, (b) cutting quality, such as quality and tolerance requirement for the workpiece that have a significant impact on product costs, (c) production rate, which has a significant impact on the economics of machining and thus essential to optimizing the machining parameters (cutting speed, feed rate, and depth of cut) to maximize the rate of production, (d) process management. Some research findings related to the study of economics in machining are described.

Davim et al. [63] performed machining of MMCs under an MQL environment at different flow rates and measured surface roughness and cutting power. It is evident from the investigation that the variation in cutting power and surface roughness at different flow rates is marginal and was found to be economical. Under the MQL condition, the performance of machining of MMC has been improved. Khan et al. [64] investigated sustainable machining of titanium alloy to assess the productivity and efficiency under hybrid CryoMQL (HCM) cooling and observed it to be better over dry and MQL machining. This technique resulted in higher tool life, minimal energy consumption, and higher productivity during machining and thus can be utilized for saving cost. Tayal et al. [65] investigated sustainable machining of Monel 400 superalloy under a dry environment and measured surface roughness, power, and cutting force. Feed rate and cutting speed were the principal factors that affected surface roughness. The reliability and economic analysis were studied for the feasibility of the cutting tool. Increase of cutting speed increases surface finish and reduces cutting force in dry machining and thus reduces power consumption. Gupta et al. [66] carried out machining of Inconel 800 alloy considering sustainability parameters as social, economic, and environmental influences. The test was performed through various sustainable cooling and lubrication conditions such as dry, vegetable oil with MQL, graphene nanofluid with vegetable oil MQL (NMQL), and liquid nitrogen (N_2). In a turning process, the types of costs include machining cost, cutting tool cost, lubrication cost, energy cost, and environmental cost.

The study concluded that liquid nitrogen application outperformed other cooling conditions and reduced total machining cost per part up to 9.3%, energy consumption by 11.3%, carbon emission by 49.17%, and tool wear by 46.6%. The proposed cooling condition improves sustainability towards environmental and economic benefits. The improvement in machining efficiency using liquid nitrogen was found to be 9.3% compared to dry machining, whereas for MQL and NMQL, it was 2.3% and 6%, respectively, as evident from Figure 5.18 [66]. Also liquid nitrogen produced 36.8%, 26.9%, and 13.2% low cost compared to dry, MQL, and NMQL techniques, as seen from Figure 5.19 [66].

For liquid nitrogen machining, a cutting speed of 100 m/min and a feed rate of 0.1 mm/rev are the recommended optimum parameters for minimization of cost and improving machining efficiency and number of parts produced. From overall comparison between cutting conditions and parameters on machining efficiency, productivity, cost, energy consumption, carbon emission, surface roughness, and tool wear in Figure 5.20 [66], it is clear that better results are obtained under the liquid N_2 cooling environment machining, followed by NMQL, MQL, and dry conditions, and that it may be considered as a sustainable cooling condition. Machinability performance along with environmental aspects are greatly improved with this approach.

Agrawal et al. [67] investigated machinability and sustainability aspects of titanium alloy through cryogenic and wet cutting environments with a possibility of adoption by the manufacturing industry. Sustainability in machining is broadly classified into process (less environmental pollution, low energy consumption, higher productivity, and worker-friendly processes) and product (longer life, better surface characteristics and dimensional accuracy) sustainability. Carbon emission is an important parameter for sustainability aspects in machining, which includes emissions through energy, material, and waste, as shown in Figure 5.21 [67].

From the analysis, the cryogenic environment performed better compared to wet machining as it reduced carbon emission by 22% at higher cutting speeds, as shown in Figure 5.22 [67]; total machining cost by 27% at higher cutting speeds, as shown in Figure 5.23 [67]; power consumption by 23.4%; surface roughness by 22.1%; and improvement in tool life by 125%. Thus sustainability has been enhanced through cryogenic cooling compared to wet machining of titanium alloy due to reduction in tool wear and power consumption, better surface quality, and less waste processing cost.

Sahoo and Sahoo [68] conducted a hard machining operation using multilayer coated carbide cutting tool and developed prediction models for responses (flank wear, Ra, and Rz). Model adequacy was checked and was found to be significant, as correlation coefficient was very high. The machining parameters were optimized through gray relational analysis and considerable reduction in responses was noticed at optimal conditions. An economic feasibility study was carried out through Gilbert's approach for the application of coated carbide insert in dry hard machining. Total machining cost per part was found to be minimum due to higher life of the cutting tool, and it thus increased savings by reducing the downtime. Therefore, it concludes the economic feasibility of utilization of carbide cutting tool in machining hardened

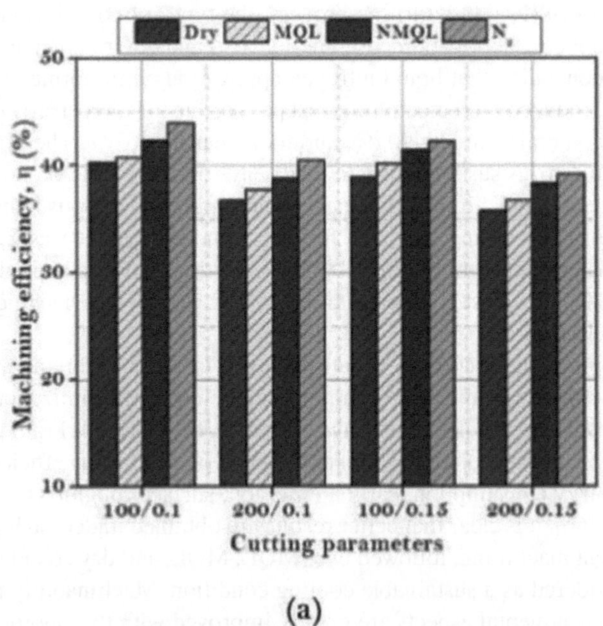

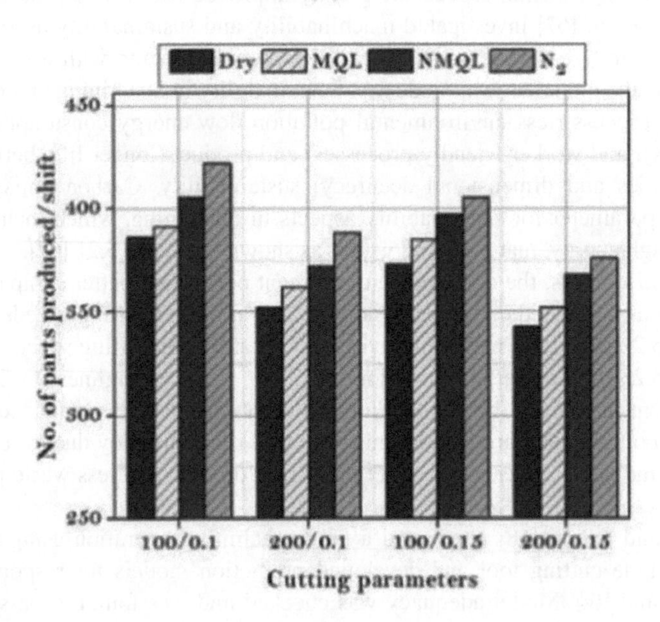

FIGURE 5.18 Impact of turning parameters (cutting speed/feed rate) and cooling/lubrication conditions on (a) machining efficiency and (b) productivity. (From Gupta, M.K. et al., *J. Clean. Prod.*, 287, 125074, 2021.)

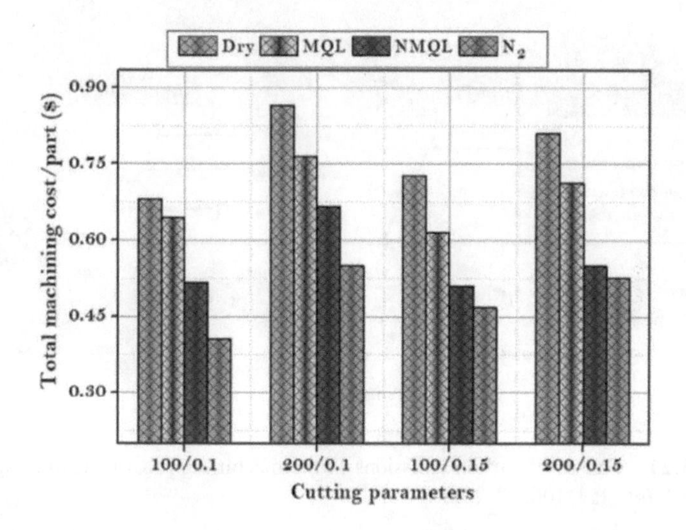

FIGURE 5.19 Impact of turning parameters (cutting speed/feed rate) and cooling/lubrication conditions on total machining cost per part. (From Gupta, M.K. et al., *J. Clean. Prod.*, 287, 125074, 2021.)

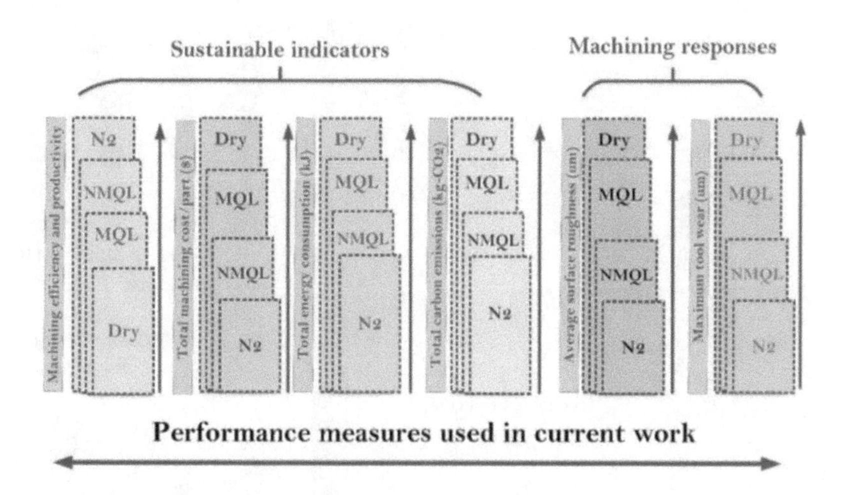

FIGURE 5.20 Overall comparison between all responses. (From Gupta, M.K. et al., *J. Clean. Prod.*, 287, 125074, 2021.)

steel. Sahoo and Sahoo [69] conducted a comparative economic feasibility study during dry machining of AISI D2 steel using uncoated and TiN-coated carbide cutting tool by estimating total machining cost per part. Cost analysis indicated 10.5 times higher machining cost per part using uncoated carbide insert compared to coated carbide during dry machining and thus yielded 90.5% cost savings. Gogna et al. [70] reviewed the fabrication techniques, effect of factors on mechanical properties, and application of biodegradable natural fiber such as jute-fiber-reinforced composites,

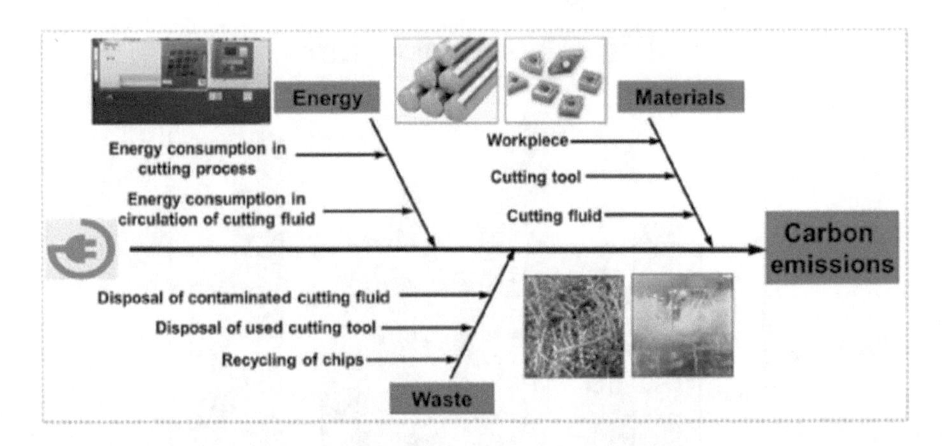

FIGURE 5.21 Causes of carbon emissions in the machining process. (From Agrawal, C. et al., *Tribol. Int.*, 153, 106597, 2021.)

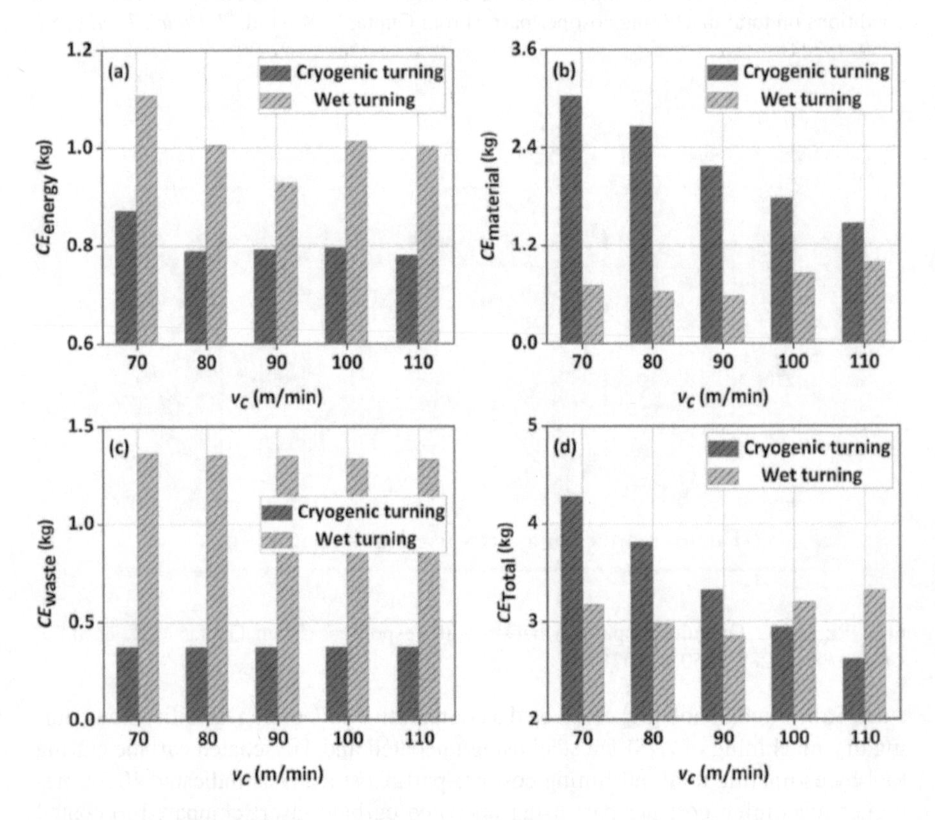

FIGURE 5.22 Comparison of carbon emissions for turning of Ti–6Al–4V under wet and cryogenic environments. (From Agrawal, C. et al., *Tribol. Int.*, 153, 106597, 2021.)

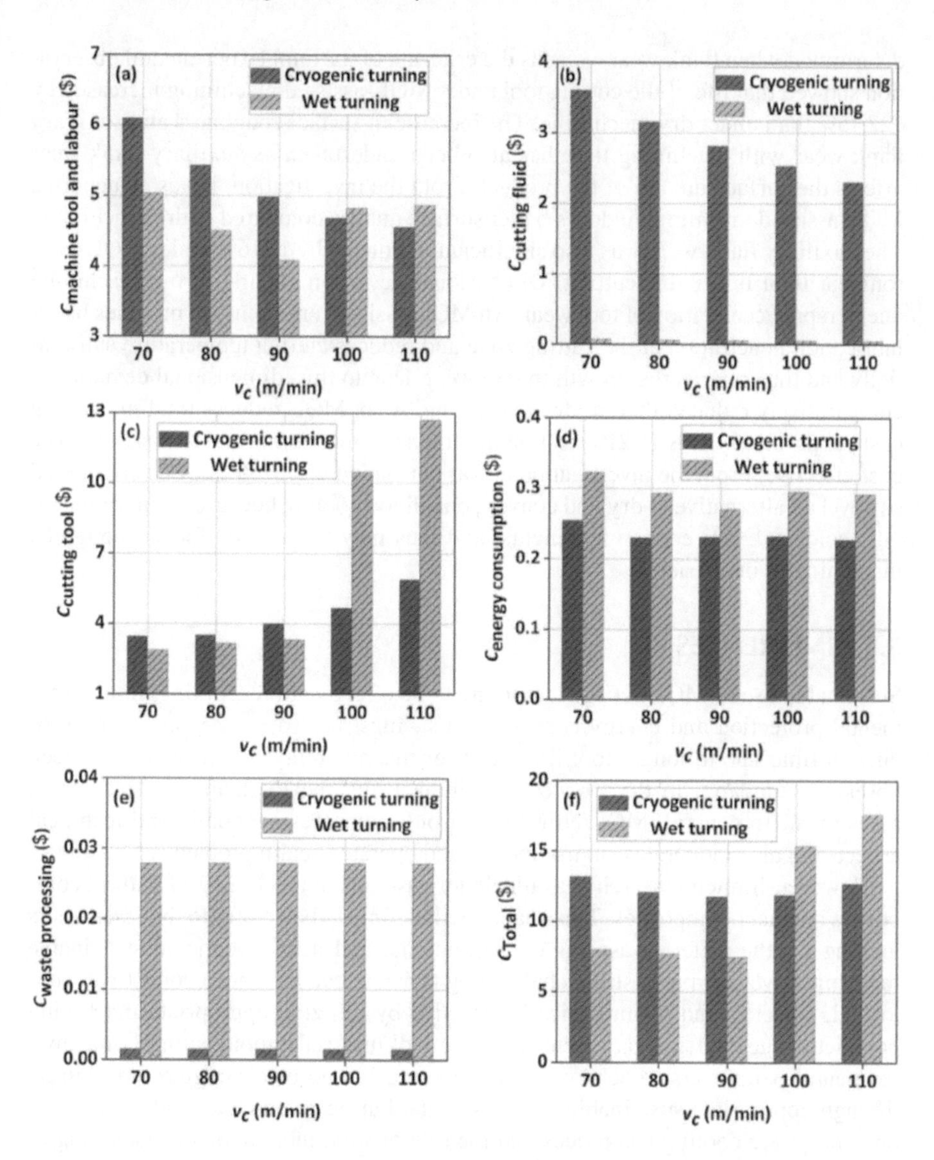

FIGURE 5.23 Comparison of cost components for turning of Ti–6Al–4V under cryogenic and wet environments. (From Agrawal, C. et al., *Tribol. Int.*, 153, 106597, 2021.)

which is successively replacing synthetic fiber and glass fiber composites. The advantages of jute fiber composite include low processing cost, i.e., economical; low energy requirement; low density; stiffness; and good mechanical properties. Das et al. [71] carried out a comparative machinability investigation under dry and MQL-assisted hard machining of AISI 4340 steel. MQL-assisted machining provides appropriate lubrication at the cutting zone and thus induced better performance compared to dry machining in terms of surface quality, flank wear, and dimensional deviation. The progression of flank wear with machining time determines the end of tool life and it is determined when flank wear exceeds the criterion of 0.2 mm. Experimental observation showed that life of the cutting tool under MQL-assisted machining increased by 122.15% than under dry machining. The increase in surface roughness and auxiliary flank wear with machining time has also been undertaken as auxiliary flank wear affects the surface quality of the product. From the investigation, it was evident that MQL-assisted machining induces better surface quality compared to dry machining. The auxiliary flank wear was also significantly reduced by 13.16% under MQL environment than under dry cutting. Dimensional deviation occurs during machining due to rapid acceleration of tool wear. An MQL-assisted environment provides better lubrication penetration at the cutting zone and reduces cutting temperature substantially and thus retards the growth in tool wear. Due to this, dimensional deviation is comparatively reduced than in dry machining. Also, MQL reduces total machining cost per part and yields 16.21% cost savings than dry machining and is observed to be cost-effective. From the investigation, it was concluded that MQL may be considered as novel an alternative to dry and conventional flood cooling because of environmental, ecological, and economic benefits, and thus may be adopted for cleaner/green machining in industries.

5.5 INFERENCES

Sustainable use of MQL/NFMQL is found to be effective in machining for environmental protection and energy consumption savings. It also reduces costs and production time due to longer tool life. Superior machinability performance has been observed compared to dry or wet machining using the Pugh matrix sustainable assessment. In general, MQL/NFMQL not only improves economic and technical aspects but also moderates environmental and operator health problems.

However, limited research is available for sustainable machining of natural composites and nanocomposites. This chapter will definitely be helpful to the researchers looking for the different aspects of conventional and nonconventional sustainable machining. Machining is still a challenging task and can be made economical, ecological, societal, and technologically effective by utilizing appropriate machining parameters, geometrical parameters, cutting tool materials, tool coatings, and environmental parameters to achieve sustainable machining or clean/green machining through appropriate sustainability assessments. Future research should focus more on appropriate cooling techniques and the role of solid lubricants for machining of difficult-to-cut materials such as nanocomposites. More research is needed for the synthesis and application of hybrid nanofluids in machining to improve machinability and sustainability performance rather than single nanoparticle inclusion. The investigations discussed here can be analyzed in detail for different types of nanofluids, nanoparticle size and shape, their concentrations, flow rate, nozzle orientation angle, spray stand-off distance, and air pressure and their effects on machining responses. Optimization of the above parameters for nanofluids is essential for effective application in machining. In addition, a life cycle assessment model may be utilized to investigate more on sustainability aspects with regard to environmental, economic, and social benefits.

REFERENCES

[1] S.K. Tamang, M. Chandrasekaran, A.K. Sahoo, Sustainable machining: an experimental investigation and optimization of machining Inconel 825 with dry and MQL Approach, *Journal of the Brazilian Society of Mechanical Sciences and Engineering*, 40, 374, 2018.

[2] A. Zein, W. Li, C. Herrmann, S. Kara, Energy efficiency measures for the design and operation of machine tools: an axiomatic approach, 18th CIRP International Conference on Life Cycle Engineering: Glocalized Solutions for Sustainability in Manufacturing, Braunschweig, Germany, 274–279, 2011. https://doi.org/10.1007/978-3-642-19692-8_48.

[3] L. Dash, S. Padhan, A. Das, S.R. Das, Machinability investigation and sustainability assessment in hard turning of AISI D3 steel with coated carbide tool under nanofluid minimum quantity lubrication-cooling condition, *Proc IMechE Part C: J Mechanical Engineering Science*, 1–33, 2021. https://doi.org/10.1177/0954406221993844.

[4] A. Shokrani, V. Dhokia, S.T. Newman, Environmentally conscious machining of difficult-to-machine materials with regard to cutting fluids, *International Journal of Machine Tools & Manufacture*, 57, 83–101, 2012.

[5] M.K. Gupta, Q. Song, Z. Liu, M. Sarikaya, M. Jamil, M. Mia, V. Kushvaha, A.K. Singla, Z. Li, Ecological, economical and technological perspectives based sustainability assessment in hybrid-cooling assisted machining of Ti-6Al-4 V alloy, *Sustainable Materials and Technologies*, 26, e00218, 2020.

[6] W. Grzesik, Nanofluid assistance in machining processes-properties, mechanisms and applications: A review, Journal of Machine Engineering, 21(2), 75–90, 2021.

[7] Shyam, M.S. Srinivas, K.K. Gajrani, A. Udayakumar, M.R. Sankar, Sustainable machining of Cf/SiC ceramic matrix composite using green cutting fluids, *Procedia CIRP*, 98, 151–156, 2021.

[8] A.S. Alaboodi, S. Sivasankaran, H.R. Ammar, Non-Traditional machining process of composite materials using renewable lubricants, *International Journal of Recent Technology and Engineering*, 9(1), 929–933, 2020.

[9] C. Kannan, R. Ramanujam, A.S.S. Balan, Machinability studies on Al 7075/BN/Al2O3 squeeze cast hybrid nanocomposite under different machining environments, *Materials and Manufacturing Processes*, 33(5), 587–595, 2018.

[10] S.K. Josyula, S.K.R. Narala, E.G. Charan, H.A. Kishawy, Sustainable machining of metal matrix composites using liquid nitrogen, *Procedia CIRP*, 40, 568–573, 2016.

[11] N.P. Hung, S.H. Yeo, B.E. Oon, Effect of cutting fluid on the machinability of metal matrix composites, *Journal of Materials Processing Technology*, 67(1–3), 157–161, 1997.

[12] S.M. Agrawal, N.G. Patil, Drilling of ceramic reinforced aluminum matrix composite under minimum quantity lubrication using bio cutting fluid, *International Journal of Innovative Technology and Exploring Engineering*, 9(4), 3172–3178, 2020.

[13] M.R. Sankar, J. Ramkumar, S. Aravindan, Machining of metal matrix composites with minimum quantity cutting fluid and flood cooling, *Advanced Materials Research*, 299–300, 1052–1055, 2011.

[14] I.P. Okokpujie, C.A. Bolu, O.S. Ohunakin, Comparative performance evaluation of TiO2, and MWCNTs nano-lubricant effects on surface roughness of AA8112 alloy during end-milling machining for sustainable manufacturing process, The International Journal of Advanced Manufacturing Technology, 108, 1473–1497, 2020.

[15] T. Lv, S. Huang, X. Hu, Y. Ma, X. Xu, Tribological and machining characteristics of a minimum quantity lubrication (MQL) technology using GO/SiO_2 hybrid nanoparticle water-based lubricants as cutting fluids, *The International Journal of Advanced Manufacturing Technology*, 96, 2931–2942, 2018.

[16] P. Sheng, M. Srinivasan, S. Kobayashi, Multi-Objective process planning in environmentally conscious manufacturing: A feature-based approach, *CIRP Annals*, 44(1), 433–437, 1995.
[17] A.A. Munoz, P. Sheng, An analytical approach for determining the environmental impact of machining processes, *Journal of Materials Processing Technology*, 53, 736–758, 1995.
[18] A.C.K. Choi, H. Kaebernick, W.H. Lai, Manufacturing processes modelling for environmental impact assessment, *Journal of Materials Processing Technology*, 70 (1–3), 231–238, 1997.
[19] Z. Jiang, H. Zhang, J.W. Sutherland, Development of an environmental performance assessment method for manufacturing process plans, *The International Journal of Advanced Manufacturing Technology*, 58, 783–790, 2012.
[20] O. Avram, I. Stroud, P. Xirouchakis, A multi-criteria decision method for sustainability assessment of the use phase of machine tool systems, *International Journal of Advanced Manufacturing Technology*, 53, 811–828, 2011.
[21] X.C. Tan, F. Liu, H.J. Cao, H. Zhang, A decision-making framework model of cutting fluid selection for green manufacturing and a case study, *Journal of Materials Processing Technology*, 129(1–3), 467–470, 2002.
[22] J. Yan, C. Feng, L. Li, Sustainability assessment of machining process based on extension theory and entropy weight approach, *The International Journal of Advanced Manufacturing Technology*, 71, 1419–1431, 2014.
[23] M. Mia, M.K. Gupta, G. Singh, G. Królczyk, D.Y. Pimenov, An approach to cleaner production for machining hardened steel using different cooling-lubrication conditions, *Journal of Cleaner Production*, 187, 1069–1081, 2018.
[24] L. Dash, S. Padhan, S.R. Das, Experimental investigations on surface integrity and chip morphology in hard tuning of AISI D3 steel under sustainable nanofuid-based minimum quantity lubrication, *Journal of the Brazilian Society of Mechanical Sciences and Engineering*, 42, 500–525, 2020.
[25] S. Padhan, A. Das, A. Santoshwar, T. R Dharmendrabhai, S.R. Das, Sustainability assessment and machinability investigation of austenitic stainless steel in finish turning with advanced ultra-hard SiAlON ceramic tool under different cutting environments, *Silicon*, 13, 119–147, 2021.
[26] A. Panda, S.R. Das, D. Dhupal, Machinability investigation and sustainability assessment in FDHT with coated ceramic tool, *Steel and Composite Structures*, 34(5), 681–698, 2020.
[27] A.T. Abbas, F. Benyahia, M.M. El Rayes, C. Pruncu, M.A. Taha, H. Hegab, Towards optimization of machining performance and sustainability aspects when turning AISI 1045 steel under different cooling and lubrication strategies, *Materials*, 12, 3023–3049, 2019.
[28] A.T Abbas, M.K. Gupta, M.S. Soliman, M. Mia, H. Hegab, M. Luqman, D.Y. Pimenov, Sustainability assessment associated with surface roughness and power consumption characteristics in nanofluid MQL-assisted turning of AISI 1045 steel, *The International Journal of Advanced Manufacturing Technology*, 105, 1311–1327, 2019.
[29] K.K. Gajrani, P.S.Suvin, S.V. Kailas, M.R. Sankar, Thermal, rheological, wettability and hard machining performance of MoS_2 and CaF_2 based minimum quantity hybrid nano-green cutting fluids, *Journal of Materials Processing Technology*, 266, 125–139, 2019.
[30] P.S. Sreejith, Machining of 6061 aluminium alloy with MQL, dry and flooded lubricant conditions, *Materials Letters*, 62(2), 276–278, 2008.

[31] S. Masoudi, M.J. Esfahani, F. Jafarian, S.A. Mirsoleimani, Comparison the effect of MQL, Wet and Dry turning on surface topography, cylindricity tolerance and sustainability, *International Journal of Precision Engineering and Manufacturing-Green Technology*, 2019. https://doi.org/10.1007/s40684-019-00042-3

[32] M. Goldberg, Improving productivity by using innovative metal cutting solutions with an emphasis on green machining, *International Journal of Machining and Machinability of Materials*, 12(1/2), 117–125, 2012.

[33] V. Upadhyay, P.K. Jain, N.K. Mehta, Machining with minimum quantity lubrication: a step towards green manufacturing, *International Journal of Machining and Machinability of Materials*, 13(4), 349–371, 2013.

[34] M.G. Faga, P.C. Priarone, M. Robiglio, L. Settineri, V. Tebaldo, Technological and sustainability implications of dry, near-dry, and wet turning of Ti-6Al-4V alloy, *International Journal of Precision Engineering and Manufacturing-Green Technology*, 4, 129–139, 2017.

[35] Y. Shokoohi, E. Khosrojerdi, B.h.R. Shiadhi, Machining and ecological effects of a new developed cutting fluid in combination with different cooling techniques on turning operation, *Journal of Cleaner Production*, 94, 330–339, 2015.

[36] S. Debnath, M.M. Reddy, Q.S. Yi, Environmental friendly cutting fluids and cooling techniques in machining: a review, *Journal of Cleaner Production*, 83, 33–47, 2014.

[37] I. Deiab, S.W. Raza, S. Pervaiz, Analysis of lubrication strategies for sustainable machining during turning of titanium Ti-6Al-4V alloy, *Procedia CIRP*, 17, 766–771, 2014.

[38] P.C. Priarone, M. Robiglio, L. Settineri, V. Tebaldo, Effectiveness of minimizing cutting fluid use when turning difficult-to-cut alloys, *Procedia CIRP*, 29, 341–346, 2015.

[39] M. Jamil, W. Zhao, N. He, M.K. Gupta, M. Sarikaya, A.M. Khan, M.R. Sanjay, S. Siengchin, D.Y. Pimenov, Sustainable milling of Ti–6Al–4V: A trade-off between energy efficiency, carbon emissions and machining characteristics under MQL and cryogenic environment, *Journal of Cleaner Production*, 281, 125374, 2021.

[40] M.K. Gupta, M. Mia, G.R. Singh, D.Y. Pimenov, M. Sarikaya, V.S. Sharma, Hybrid cooling-lubrication strategies to improve surface topography and tool wear in sustainable turning of Al 7075-T6 alloy, *The International Journal of Advanced Manufacturing Technology*, 101, 55–69, 2019.

[41] O. Pereira, J.E.M. Alfonso, A. Rodríguez, A. Calleja, A.F. Valdivielso, L.N.L. de Lacalle, Sustainability analysis of lubricant oils for minimum quantity lubrication based on their tribo-rheological performance, *Journal of Cleaner Production*, 164, 1419–1429, 2017.

[42] R. Kumar, A.K. Sahoo, P.C. Mishra, R.K. Das, Measurement and machinability study under environmentally conscious spray impingement cooling assisted machining, *Measurement*, 135, 913–927, 2019.

[43] R. Kumar, A.K. Sahoo, P.C. Mishra, R.K. Das, S. Roy, ANN modeling of cutting performances in spray cooling assisted hard turning, *Materials Today: Proceedings*, 5(9), 18482–18488, 2018.

[44] P.C. Mishra, D.K. Das, M. Ukamanal, B.C. Routara, A.K. Sahoo, Multi-response optimization of process parameters using Taguchi method and grey relational analysis during turning AA 7075/SiC composite in dry and spray cooling environments, *International Journal of Industrial Engineering Computations*, 6(4), 445–456, 2015.

[45] R. Kumar, A.K. Sahoo, R.K. Das, A. Panda, P.C. Mishra, Modelling of flank wear, surface roughness and cutting temperature in sustainable hard turning of AISI D2 steel, *Procedia Manufacturing*, 20, 406–413, 2018.

[46] S.K. Sahu, P.C. Mishra, K. Orra, A.K. Sahoo, Performance assessment in hard turning of AISI 1015 steel under spray impingement cooling and dry environment, *Proceedings of the Institution of Mechanical Engineers, Part B: Journal of Engineering Manufacture*, 229(2), 251–265, 2015.
[47] R.K. Das, A.K.Sahoo, P.C. Mishra, R. Kumar, A. Panda, Comparative machinability performance of heat treated 4340 Steel under dry and minimum quantity lubrication surroundings, *Procedia Manufacturing*, 20, 377–385, 2018.
[48] R. Kumar, A.K. Sahoo, P.C. Mishra, R.K. Das, Influence of Al_2O_3 and TiO_2 nanofluid on hard turning performance, *The International Journal of Advanced Manufacturing Technology*, 106, 2265–2280, 2020.
[49] S. Roy, R. Kumar, A.K. Sahoo, R.K. Das, A brief review on effects of conventional and nano particle based machining fluid on machining performance of minimum quantity lubrication machining, *Materials Today: Proceedings*, 18(7), 5421–5431, 2019.
[50] R.K. Das, R. Kumar, G. Sarkar, S. Sahoo, A.K. Sahoo, P.C. Mishra, Comparative machining performance of hardened AISI 4340 Steel under dry and minimum quantity lubrication environments, *Materials Today: Proceedings*, 5(11), 24898–24906, 2018.
[51] M. Ukamanal, P.C. Mishra, A.K. Sahoo, Temperature distribution during AISI 316 steel turning under TiO_2-water based nanofluid spray environments, *Materials Today: Proceedings*, 5(9), 20741–20749, 2018.
[52] M. Ukamanal, P.C. Mishra, A.K. Sahoo, Effects of spray cooling process parameters on machining performance AISI 316 Steel: a novel experimental technique, *Experimental Techniques*, 44, 19–36, 2020.
[53] S. Dambhare, S. Deshmukh, A. Borade, A. Digalwar, M. Phate, Sustainability issues in turning process: A study in Indian Machining Industry, *Procedia CIRP*, 26, 379–384, 2015.
[54] A. Das, S.K. Patel, B.B. Biswal, S.R. Das, Performance evaluation of aluminium oxide nano particles in cutting fluid with minimum quantity lubrication technique in turning of hardened AISI 4340 alloy steel, *Scientia Iranica B*, 27(6), 2838–2852, 2020.
[55] Y. Iskandar, A. Tendolkar, M.H. Attia, P. Hendrick, A. Damir, C. Diakodimitris, Flow visualization and characterization for optimized MQL machining of composites, *CIRP Annals*, 63(1), 77–80, 2014.
[56] M.A.ul. Haq, S. Hussain, M.A. Ali, M.U. Farooq, N.A. Mufti, C.I. Pruncud, A. Wasim, Evaluating the effects of nano-fluids based MQL milling of IN718 associated to sustainable productions, *Journal of Cleaner Production*, 310, 127463, 2021.
[57] S. Qu, Y. Gong, Y. Yang, W. Wang, C. Liang, B. Han, An investigation of carbon nanofluid minimum quantity lubrication for grinding unidirectional carbon fibre-reinforced ceramic matrix composites, *Journal of Cleaner Production*, 249, 119353, 2020.
[58] K. Giasin, S. Ayvar-Soberanis, A. Hodzic, Evaluation of cryogenic cooling and minimum quantity lubrication effects on machining GLARE laminates using design of experiments, *Journal of Cleaner Production*, 135, 533–548, 2016.
[59] K. Giasin, S. Ayvar-Soberanis, A. Hodzic, The effects of minimum quantity lubrication and cryogenic liquid nitrogen cooling on drilled hole quality in GLARE fibre metal laminates, *Materials & Design*, 89, 996–1006, 2016.
[60] N. Bhanot, P.V. Rao, S.G. Deshmukh, Sustainable manufacturing: An interaction analysis for machining parameters using Graph Theory, *Procedia – Social and Behavioral Sciences*, 189, 57–63, 2015.
[61] R. Somaashekaraiah, P.S. Suvin, D.P. Gnanadhas, S.V. Kailas, D. Chakravortty, Eco-friendly, non-toxic cutting fluid for sustainable manufacturing and machining processes, *Tribology Online*, 11(5), 556–567, 2016.
[62] S. Padhan, L. Dash, S.K. Behera, S.R. Das, Modeling and optimization of power consumption for economic analysis, energy-saving carbon footprint analysis, and sustainability assessment in finish hard turning under graphene nanoparticle–assisted minimum quantity lubrication, *Process Integration and Optimization for Sustainability*, 4, 445–463, 2020.

[63] J.P. Davim, P.S. Sreejith, J. Silva, Some studies about machining of MMC's by MQL(Minimum Quantity of Lubricant) conditions, *Advanced Composites Letters*, 18(1), 21–23, 2009.
[64] A.M. Khan, N. He, L. Li, W. Zhao, Analysis of productivity and machining efficiency in sustainable machining of titanium alloy, *Procedia Manufacturing*, 43, 111–118, 2020.
[65] A. Tayal, N.S. Kalsi, M.K. Gupta, A.G. Collado, M. Sarikaya, Reliability and economic analysis in sustainable machining of Monel 400 alloy, *Proceedings of IMechE Part C: Journal of Mechanical Engineering Science*, 2021, https://doi.org/10.1177/0954406220986818.
[66] M.K.Gupta, Q. Song, Z. Liu, M. Sarikaya, M. Jamil, M. Mia, A.K. Singla, A.M. Khan, N. Khanna, D.Y. Pimenov, Environment and economic burden of sustainable cooling/lubrication methods in machining of Inconel-800, *Journal of Cleaner Production*, 287, 125074, 2021.
[67] C. Agrawal, J. Wadhwa, A. Pitroda, C.I. Pruncu, M. Sarikaya, N. Khanna, Comprehensive analysis of tool wear, tool life, surface roughness, costing and carbon emissions in turning Ti–6Al–4V titanium alloy: Cryogenic versus wet machining, *Tribology International*, 153, 106597, 2021.
[68] A.K. Sahoo, B. Sahoo, Performance studies of multilayer hard surface coatings (TiN/TiCN/Al_2O_3/TiN) of indexable carbide inserts in hard machining: Part-II (RSM, grey relational and techno economical approach), *Measurement*, 46(8), 2868–2884, 2013.
[69] A.K. Sahoo, B. Sahoo, A comparative study on performance of multilayer coated and uncoated carbide inserts when turning AISI D2 steel under dry environment, *Measurement*, 46(8), 2695–2704, 2013.
[70] R. Kumar Gogna, A.K. Anurag, A. Panda Sahoo, A comprehensive review on Jute Fiber reinforced composites, lecture notes in mechanical engineering, *Advances in Industrial and Production Engineering*, 459–467, 2019.
[71] R.K. Das, A.K. Sahoo, R. Kumar, S. Roy, P.C. Mishra, T. Mohanty, MQL assisted cleaner machining using PVD TiAlN coated carbide insert: Comparative assessment, *Indian Journal of Engineering & Materials Sciences*, 26, 311–325, 2019.

[illegible] P. [illegible], T. [illegible] [illegible] [illegible] [illegible] of MMCs [illegible] Titanium [illegible] of [illegible] composites [illegible] 8, [illegible] 2001.

[64] A. [illegible], N. [illegible], E. [illegible] [illegible] [illegible] [illegible] machining [illegible] and [illegible] machining of titanium alloy [illegible] 2014.

[65] A. [illegible] N. S. [illegible], M. [illegible] [illegible] [illegible] [illegible] [illegible] [illegible] [illegible] [illegible] [illegible] [illegible] [illegible] [illegible] [illegible] 2017. https://doi.org/10.1177/[illegible]

[66] [illegible] [illegible] [illegible] [illegible] [illegible] M. [illegible] [illegible] Singh, M. [illegible] N. [illegible] [illegible] [illegible] [illegible] and [illegible] [illegible] [illegible] [illegible] [illegible] and [illegible] [illegible] [illegible] [illegible] [illegible] 2021.

[67] [illegible] [illegible] W. [illegible] [illegible] [illegible] R. [illegible] [illegible] Comparative analysis of tool wear and the surface [illegible] [illegible] [illegible] [illegible] [illegible] [illegible] [illegible] [illegible] [illegible] [illegible] [illegible] [illegible] [illegible]

[68] A. [illegible] B. [illegible] [illegible] studies in [illegible] [illegible] [illegible] [illegible] [illegible] [illegible] carbide [illegible] [illegible] [illegible] Part [illegible] [illegible] and [illegible] [illegible] [illegible] [illegible] 2021 [illegible]

[69] [illegible] B. [illegible] A. [illegible] study on performance of [illegible] [illegible] [illegible] [illegible] [illegible] [illegible] [illegible] [illegible] [illegible] [illegible] [illegible] 2016 [illegible]

[70] [illegible] A. S. [illegible] A. [illegible] [illegible] [illegible] Review on fiber reinforced composites [illegible] [illegible] in [illegible] [illegible] [illegible] [illegible] [illegible] [illegible] [illegible]

[71] [illegible] B. [illegible] [illegible] [illegible] [illegible] [illegible] [illegible] [illegible] [illegible] [illegible] using PVD [illegible] [illegible] [illegible] [illegible] [illegible] [illegible] [illegible] [illegible] [illegible] 2015.

Index